Alsatian Shepalute's

A New Breed for a New Millennium

By

Lois Denny

First published by AuthorHouse 08/13/04

ISBN: 1-4184-6998-X (e-book)
ISBN: 1-4184-3922-3 (Paperback)
ISBN: 1-4184-3921-5 (Dust Jacket)

Library of Congress Control Number: 2004092461

This book is printed on acid free paper.

Printed in the United States of America
Bloomington, Indiana

Bach von Razz Berenty (Five-Months old)

Bolo 13 years old summer Colorado

I wish to thank you for purchasing this book! My mother thanks you; my husband thanks you and most of all my dogs' thank you!

100 % of the proceeds from this book will go directly towards the further enhancement of this great breed. The care, feeding, boarding and recording of the dogs as well as the adoption process necessary to provide each pups with a proper loving and devoted home.

The Alsatian Shepalute

To all those who believed in me, who loved me through it all, and who encouraged me when I was at my lowest hours.

To all those who have bought, owned and loved these dogs and who have encouraged me never to give up.

To my mom who knows nothing about the dog world, the different dog breeds, or doggy language. May this book help you to understand my passion.

And last but not least,
To my husband who went nights without dinner and took second place
To the Alsatian Shepalutes in my life
Thank-you.

Table of Contents

Introduction

The Alsatian Shepalute book grew out of the need for the organization of such content in book form. Though this book is only general information that has been compiled from all my research, studies, breeding's and my works, I do realize it is not enough for the very interested amongst you. I promise all of you that are still waiting patiently, that the next book will go further into the generations (with more pictures) of the complete development of this breed. The next book will include all the pure-bred registered breeds (A.K.C, U.K.C, S.V and C.K.C) from the first generations to the newest generations, as I have them recorded and photographed with charts on all of the dogs. That might be a little bit boring for most of the public, so for now this book shall simply have to be enough to introduce this great breed of the new millennium!

With this book, I introduce the Alsatian Shepalute to the world! Is the world ready for such a dog? Will they take care of the complex structure and intuitive nature of this magnificent breed?

It is my wish that only the great dog lovers of the world get involved with this breed for therein lies the true test of mankind. This breed is not a fashion statement, it is an animal. As a companion dog, its only function is to be a part of your family, to share memories and to kiss the faces of your children. Do not get me wrong, this dog will do as much for any human as any dog before it, as this breed has come from the best!

This dog was taken out of the fashion and greed of the large conglomerates and their breed clubs and has been placed in the gentle hands of the loving American families. My only hope and prayer is that it will forever stay there! It is up to you, the owners, to get involved in the Alsatian Shepalute Breed Club in order that this breed may not regress. By sharing with other Alsatian owners, the characteristics of the breed and their standards, you will be helping to keep this breed from deteriorating as so many other breeds have done throughout the years.

When I was a child of thirteen, I asked my higher power, my God, a question of depth, "why was I born?" I expected a large complicated story or at least the words "it is too deep for you to understand." Instead a peace came over me and an awakening, as my mind told me the answer and God spoke, "to be happy." I am glad to have that answer for that simple answer has helped me through so much in life!

So, I say to you with open arms, "Here is my love, my life, my compassion and all that I am. Here is The Alsatian Shepalute, go and be happy!"

L. E. Schwarz

Quotes

Tye Cobb: The desire for glory is not a sin, the world needs heroes.

Chief Seattle —- All things are connected.
Whatever befalls the earth
Befalls the sons of the earth.
Man did not weave the web of life.
He is merely a strand in it.
Whatever he does to the web,
He does to himself.

Lois Schwarz —- It's in the Genes

Forward
by Jennifer Kingsley

I know of the trials that you endured to get where you are. I know that you almost lost them all. Packer, Betty and Bach were your only links back to all the dogs that you loved so much throughout the years. Your life is truly a legacy and when you do pass on after many years from now, I'm sure there will be books written about you and all that you did for the love of life. So many people do not aspire the way you do. They live their lives from day to day doing the same monotonous things that have so little meaning in reality. What will they leave for others to cherish? What impact did they have on others in this world? I believe in your dream and in your dogs. They are everything that is good in life. They are companions to those in need. I wish the whole world could see how wonderful these dogs are. Just the love that emanates from their eyes. The intelligence they possess. The wisdom of knowing your every movement and wanting to please you always. They are the ultimate friends. A friendship that no human can match. The love this dog has for its owner is comparable to the love Jesus shared with the world. They love unconditionally. They are patient, kind, honest, faithful, and trusting. They give themselves to you completely. What a love that is! If we humans could only give half of the love these dogs give, we would all be better people for it. Everyone should own a dog in their lifetime, just to have that awesome love that they give. Especially the Alsatian Shepalute. He is beauty and grace. Love and loyalty. Friend and companion.

You had a dream so long ago, through so much pain and suffering, but also joy and happiness. Your road has taken you through so much growth. People do not have the background that you have to claim what you claim. You have dedicated your WHOLE life to this dream. Where many would give up or never start, you have gone far beyond.

I'm sure people thought that you were crazy for starting this breed, and I'm sure people thought that you would not last and would give up sooner or later. Even now they may think that. But, you've endured. This breed is a true expression of you. All that you want to be and aspire to be are in these dogs. NEVER stop your dream.

A gift has been given to you in these dogs. I can see your vision. I can see your love. I can see all that you've given to your life.

If you believe in something, then you must live it to the fullest. Bravo. I say Fantastic! Unbelievable! Tremendous! To dream, what a gift. To realize your dream, what an honor.

I love you so much. Not for what you've accomplished in your life, although that is great. But, for you. Just for all the love that you have and give freely. You are the most loving person I know. You are true. Your feelings are always true.

Jennifer Kingsley
(President of the Alsatian Shepalute club)

Chapter 1

The History of the breed

The Creator of the breed

In Panama, on October 27, 1953, a daughter was born to Robert and Jane Denny whom they named Lois Elaine Denny. Seven years later, the Denny family moved to Oxnard, California where Mr. Denny worked at the Point Mugu Naval Air base until he retired.

Lois had a genuine gift with animals and would prefer them to most people in her later years. While Lois was growing up in Oxnard, she joined the brownie, juniors, cadets and senior scouts. She enjoyed this comradery so much that she then went on to be a brownie leader herself.

She carried her love of animals with her always. Her main interest as a child was in the breeding and raising of her guinea pigs, and mice which gave her a peace and happiness that showed throughout her soul. Soon she took over the family's pigeons and had a hand in the rabbits, ducks and chickens. She trained her ducks and chickens on cat leashes going for short walks throughout the neighborhood. She also built a small maze, as she was inquisitive about what she had read on food rewards and memory, along with the salivation of glands and the dinner bell. But her real love would always remain with the canines.

By the time Lois was ten she had read all books of such interests that were contained in the libraries near her home.

Lois kept records of the coat colors, coat density, eye shapes, mutations, and albinism that occurred within her breeding's. She received numerous awards while in scout's within these fields. Then there were the dog club ribbons and the Girl Scout dog show event's that she would put on for the community.

Lois joined the local dog training club's, but would see the political side and drop out as she also dropped out of the beauty contest for Miss Oxnard even though she had won second runner up in the Miss Wendy Ward competition. She realized early on in life that she did not need

1

such recognition's that stripped her of her intense loyalty to others and to herself by substituting truth for acceptance. If she had to play such games to win the judge's votes, then she wanted no part of the game that others were so willing to play.

Lois would train the dogs she owned by herself without the lies of political correctness spewing out of the mouth's of the soon to be animal right groups in the making. Lois would also see the decline of the rodeos; circus's and pet shops in the near future.

Her favorite pet shop was an old converted barn on Oxnard Boulevard called "Valentine's Feed and Seed". Valentine's was a family run business where there were a variety of animals, such as: goats, turkeys, pigeons, rabbits, and pigs. Some of Lois's fondest memories of her childhood and the smells of freshly mowed hay and feed were there, in that barn. Those smells would be an ever-present reminder of a past that would fade away all too quickly.

When did the Idea of this great breed of dog enter her mind? The seed's of this new breed were planted in 1967 when a large mixed puppy followed her home. The Idea of the new breed "Alsatian Shepalute" would not come together for her until many dogs, breeds, and training's later.

Lois trained thousands of dogs for people and during this time she tried very hard to educate the owners on the many different dog breeds and their personalities. She would tell the truth about canines and how they really learned. She would suggest the easiest ways to train them. What she found out was that many people did not want to hear the truth. These people were not from the farms and mountains. They had a living that would erase the past of the real world and replace it instead with comfortable images of what the world was to them.

Owners of Labrador and Golden Retriever's of the 1970-80's wanted a seeing-eye dog that was trained to do everything they imagined when they first brought the puppy home. These dogs then came to Lois at that mischievous age of eight-months-old and totally wild! These were the bird dogs! Labradors with extensive lineages going back to the days when folks hunted birds for food. What were the owners expecting? Gee wiz, if the dogs didn't possess those bird hunting traits, they wouldn't be bird dogs!

Soon to be puppy owners didn't know what they wanted (but thought they did) as they came to Lois for advice. New puppy buyers may have read up on a breed and looked at the pictures of the finely groomed pets that covered the pages of breed books. Then they would decide that what they really wanted was a blue eyed Siberian or a puffy white-coated Samoyed husky!

When one has over 30 years of experience listening and coming in contact with such disasters as choosing a pup on looks alone, one begins to know what the public is looking for. Lois then decided that she just had to create that perfect companion dog! Those dogs that everyone wanted! She knew she could do it! She knew it would take a long time. She was a dreamer, a lover of that easily trained intelligent dog, but where is he?

Lois and her fifth and sixth generation cottony coated solid chocolate American cockers 1976.

In the Beginning

The beginning actually happened with home study education of the many different breeds throughout the world and the educational knowledge of the dog clubs, associations, and dog registry outfits. My library of dog books and encyclopedias of dog breeds grew. If there was a canine in a far corner of a remote island, I read up on it.

Sometime during the early 70's, I developed sixth generation solid chocolate American Cocker Spaniels. I tried the A.K.C ring, but was once again disappointed with the disregard for the standards of the breed by their judges. Who needs them if they do not even uphold the standards of the breeds? Do they even know the great consequences of their actions?

My sense of humor took me into the rings with a clean and brushed, (but not clipped) purebred registered American Cocker. They didn't want to see that full-coated hairy cocker that couldn't get a bird if his life depended on it! And why not? Didn't they like what they created? Or did the grooming industry's pockets grow fat with this particular coat? What was happening to the true American Cocker Spaniel? Where have they all gone? And what was happening to the dog world?

I studied the rules of many Kennel Clubs to find the answers. Then, I switched to registering my cockers from the American Kennel Club (A.K.C) to the United Kennel Club (U.K.C). At least they required a drawing of the markings on the dogs. I then found out that the public knew little about the different registration clubs. They didn't want U.K.C papers! Who were those guys anyways?

Mindy Von Edelhause 1984

All of my many relationships to different breeds led me back time and time again to the German Shepherd Dog. In most of these dogs were the brains and the intelligence to be able to make a great companion dog, if only they weren't so hyper!

The German Shepherd Dog was wanted by many of my customers and friends. Everyone wanted a loyal intelligent companion dog that could protect the children and the family. But what they didn't want was that hyper, ready to go (with the energy it takes to patrol a Nazi borderline) type of dog. These shepherd owners would continue to call me to have me place their dogs in other people's homes as they were tearing up the backyards or barking consistently.

I knew that this new breed of dog I envisioned would need to be bred with a complementary breed of dog that was not aggressive, that didn't bark much and that loved people. A breed with a great body! I had felt that the shepherds had fallen short in this area. Comparing photos of the original shepherds to the shepherds of today, I felt German Shepherd breeders had gone too far away from their true breedings.

This new breed that I would create would have to be a large dog. And then the look came to me, the wolf! Wolf hybrids were starting to grab the public's attention. Perhaps I could satisfy the love of the look of a wolf and the need to own a wolf with this new companion dog that would only "look" like the wolf! Maybe then folks might stop breeding wild wolves

5

with dogs and sending these hybrids to early graves because wolf/dog crosses could not fit into a domesticated family life.

I went to the information center in my brain and calculated that the Alaskan Malamute bred to the German Shepherd Dog should do the trick. Next, and the most important factor in making this new breed of dog, would be the selective hand picking of the desired characteristics and the so important breed standards. I went to the drawing board and wrote up the standards of this new breed. What would it look like? Act like? And what were the most important qualities that would keep the dog in good health? To these very important qualities were awarded certain points that were incorporated into the standards to protect the future generations.

Hyper ness conflicted with intelligence. For the public to be able to train this dog, the dog itself would have to be attentive, watchful and not bounce around all over the place. I found that too many dogs were taken to the pound because new owners could not devote the time necessary to train their animals to stop digging or barking. These were unwanted traits. I knew that this new breed could be selectively bred not to dig or bark.

This breed just had to resemble the wolf in my mind. Not the large coyote or the skinny wolves that looked like breed crosses back to a coyote, but perhaps closer to the dire wolf of days gone by.

Now the name had to be chosen. The first name that came to me was the North American Shepalute. Combining both the German Shepherd and Malamute names together. Then, in the year 2000, I settled on the Alsatian Shepalute because I had read that the word Alsatian meant wolf dog to many folks.

The meaning of the word Alsatian
In my opinion, after Capt. Max bred the three types of sheep herding dogs together from the countryside's, hillsides and farmlands around Germany, he then sent his dogs out across the countries closest to home by selling those (German) Sheep herding dogs.

Now the public had never heard of these dogs and when they saw them they really thought that these dogs were hybrid wolves. (Back then it was not a popular thing to be called a hybrid wolf). If a farmer saw one of these wolf looking dogs, well I am afraid he'd have to kill it before it killed his stock. Therefore the name Alsatian was a bad name, a slang name that had attached itself to Capt. Max's dogs! He wanted the world to accept his dogs as pure of breed, not as a mix of the wild lobo. That was kind of a slap in the face to him and to the proud owners of these dogs, not to mention that the country of Germany wasn't even getting the honor of origin on this dog. The group of folks that had an interest in these dogs had to do something and fast before the name "Alsatian" stuck. The breeders and owners came together and registered these dogs in England as the "German Shepherd Dog" so that everyone would stop referring them to a mixed wild mutt. Now they had a certificate from the world of dogs! They had the official name from a growing association. (The first dog show was held in England in 1859. It was there that the dog game was born).

Now-a-day's the name Alsatian does not bring forth such disgrace. As a matter of fact, many owners of German Shepherds are proud of the name Alsatian! Some German Shepherd breeders would like to keep that name alive so they started calling all the sable gray pups in their litters "Alsatians." (Some may have thought that the word Alsatian represented the wolf gray or the sable color of the gray German Shepherd Dog). The name Alsatian does not nor has it ever represented the breed "The German Shepherd Dog". To say that it does is incorrect.

I looked the word Alsatian up in the Encyclopedia Britannica and it does refer me to the German Shepherd Dog. I thought I might see a country attached to such a name, but it was not there. I am sure this attachment to the word Alsatian does not please Germany. Not only that, it is wrong.

Some German Shepherd breeders used the name "Alsatian" to refer back to the old'n days, so to speak. Some used the name to impress others that they knew more about the history of the breed than the mere public did. And some German Shepherd breeders use the name Alsatian in order to get their advertising in the "A" listings.

I had learned from some shepherd folks that the word Alsatian meant "wolf dog looking" or a hybrid wolf dog mix. So I did some research and I found out that there is a piece of land called Alsace. Alsace was a part of Germany for several centuries but was given to France in 1648. Germany's victory in the Franco-Prussian war brought it the entire town of Alsace. The bloodiest battles were fought around and through-out that small town. Back and forth this town went from France to Germany and then back again to France.

Alsace lies in the crossroads of trans-European travel so it might be that these dogs were brought through the town into the European countries.

Perhaps this German Shepherd Dog might have had a serious battle of names had the Verien fur Deutsche Schaferhunde not stepped in immediately to reclaim this sheep herding dog as the "German" Shepherd dog. The Verien fur Deutsche Schaferhunde (or VS.) was founded in 1899.

Perhaps the folks in England called the dog's Alsatian sheep herding dogs because they looked like the dogs that herded sheep in the Northeast provinces of France. Maybe Capt. Max bred one of those western German sheep herding dogs into his line and his dogs resembled the French sheep herding dogs? Some say that because of Hitler the world turned their backs on the use of the word "German" and anything having to do with Germany. Therefore the world renamed all dogs using the word "German" in the description of the dog, to anything but. Since Alsace was the closest country of origin of these dogs, the folks directly in contact with these dogs may have began referring to them as the Alsatian sheep dogs.

Oh yeah, where was I? I can get a bit carried away. Going back to the reason I chose the name Alsatian Shepalute… it was when my dogs had finally gotten rid of the dominate black saddles and obtained those recessive yellow eyes that the name Alsatian Shepalutes just had to be acquired, as they did so look like a wolf! It just seemed to fit.

First Litter of Malamute x German Shepherd mix 1988

The First Generations

On February 2, 1988 the first litter and the first generation (f-1) North American Shepalute's were born in the city of Oxnard. The registered name for the German Shepard Dam was "Kingsley's Swanny River." The Alaskan Malamute Sire was "Buddy of Cold Spring's."

Swanny possessed both German Schutzhund lines and American show lines within her pedigree. She had just had a litter of pups down in Georgia and I went to look at them. I saw the brother of the dam and the sire of the pups. I also saw the father of the bitch "Swanny" and asked the family if I could purchase the flea-ridden mother for $75.00. The deal was made and Swanny came home to California to meet Buddy the Alaskan Malamute!

Buddy was a large short-coated malamute with a friendly, lovable and clownish personality. Buddy loved to go fishing and could look out of the driver's side of the little pick-up truck while his tail wagged out of the passenger's side window. Buddy weighed 145 pounds at 15 months old.

I believed that the German Shepherd Dog and the Alaskan Malamute, when bred together, would bring me the closest to the standards I had written down. While the selective breeding of these breeds brought me closer to the mental character that I had envisioned, the physical outcome was still lacking.

Years later the English Mastiff was breed into the lines. This gave the breed the massive body conformation.

As with the shepherd x malamute cross, the introduction of the English Mastiff also brought along its faults. The pups had to once again be hand selected for boldness as shy pups are always dominant. The physical looks of the new breed once again had to take the backseat to its inner beauty. It took me over 15 years to get the final look of these dogs.

Only through real experience and knowledge on this subject can one begin to understand the complexity and the importance in the selection that is necessary to obtain the desired results. This is very important!

The selections of parents who will give their genes to be carried on into new generations cannot be chosen on looks alone. A breeder must be able to choose the right pup out of a litter. This chosen pup must carry the genes that will bring that dominant gene into the new breed in the desired consistency. I had to look deeper into the pup and figure out if that pup would breed back the desired traits.

Along with selecting pups with the right character and temperament a breeder must choose a pup that will not grow up to have physical deformities. This is where one must know the dogs being bred. Not just the immediate upline, but the many generations of aunts and uncles and grandparents that have all contributed to that one pup.

Today After 26 years of breeding, researching and recording the vital statistics of the litters and the logging of numerous pages of Shepalute owners, The Alsatian Shepalute has finally arrived! A dog that fits the standards of the breed that I had envisioned so many years ago! A breed of dog with the loyalty and compassion of the German Shepherds without that working energy or that shepherd whine. A dog that doesn't dig and run the fence line as this breed was not intended to be a working breed but a companion dog.

Packer Von Der Schwarz (2001) Five years old.
Previous page: Trixy Von Der Schwarz daughter of Packer. Five months old. 2003

Author's sketch from an 1809 print

The Northern Breeds

The information that I researched for this heading "The Northern Breeds" came mostly from the "Standard Guide to Purebred Dog's" written in 1977 and "The Alaskan Malamute" co-written by Eva Seely herself along with many other research books.

Just because a subject is written down on paper doesn't mean that the person who wrote it knew what they were writing about or told the truth. Maybe it was the truth as they saw it. And maybe that wasn't really the truth at all. Maybe they just felt it had to be the way it really was. I mean do you really believe everything you read? Gee, you know half the time the newspapers themselves get something wrong. OK, almost everything wrong.

Neither, of course, should you assume that I know what I am talking about! I challenge you to do your homework! Find the truth! Research books and encyclopedias before the 1800's. Maybe you will see something different than I have. What I have come to learn is that the breed registry clubs will accept any so called made-up breeds. They will even accept breeds that one claim are very old breeds regardless of proof. They do not care where the dog came from nor do they care about the breed itself. It's big business in all its glory!

To read more on the subject of breed registry clubs go to chapter 3 of this book, which is titled: Creating New Breeds of Dogs. Then look under the heading of "Breed Registration Clubs."

Kiak of Cold Springs Great Dam of the breed Alsatian Shepalutes

The Alaskan Malamute —- One can put the Alaskan Malamute and the Siberian Husky together in a package and call them the true handmade dogs of the North. Now I don't care if someone wishes to create a breed as Eva Seeley and her friends did, but what bugs me is that the public is led to believe that these Alaskan Malamute dogs are the true husky dogs of the North. (The only folks who know the real truth are those who created these breeds). If anyone would like proof, read up on any information about the Northern dogs that pulled the sledges for the Eskimos before the influx of civilized man and those gold seekers. The secret here is to read only the information on such dogs before the influx of the gold seekers and that would be about 1700 - 1800. The reason being is that those who wrote about the "true Northern breeds" went up to Alaska after the introduction of the dogs from the states! What does that tell you?

Selling dogs was big business up there at that time and there were not many dogs to sell unless one brought them over. That's why so many profitable individuals put on their thinking caps and shipped dogs into Alaska. The gold miners thought the dogs they saw were indeed the dogs of the Northern country when in fact they were not. If you found anything written on those true Northern dogs it may have read something like this:

The Northern dogs were a mutt with no breeding program. Some of the dogs ears were up and some down. These Eskimo dogs worked for the Eskimo people and if they couldn't or didn't do the jobs expected of them they were eaten. It was a hard life of survival for the Eskimo Indians. Their dogs were a useful tool in that survival. These Northern dogs had several roles to play. But the most important was that of a sledge hauler.

These dogs had long muzzles and lots of energy! They had a tail that tightly curled up over the back. They were a true spitz type of dog. These dogs would go for long periods of time out in the cold with little or no food. These Northern dogs were not identified by breed, as they were not a breed. They were just dogs that helped the Indians survive. In asking the name of the dog, the Alaskan Indian would use the word "Kingmik" or "Qimmiq." (That means "dog" in English.)

There were several Indian tribes throughout the North. The Malamute Indians had dogs that people would refer to as the "malamute" dogs. The Husky Indian tribes had dogs that people would refer to as the "Husky" dogs. Some naturalists just called the dogs "sledge dogs" or "Alaskan dogs" or whatever name they could think of to communicate to others what the heck they were talking about.

Robert Zoiler was around at the time Eva Seely introduced this new breed to the American Kennel Club and he has written some things about the Alaskan Malamute and the dogs of the North. He was also a friend of Eva's and a breeder of the Alaskan Malamute when the breed had just begun.

The new Northern breeds
Now, Eva Seely knew most of the folks who worked in the American Kennel Club and together they got the Alaskan Malamute those "pedigrees" with the name **Alaskan Malamute** written across the top of it. That was the beginning of this breed!

The Next thing one would need to do was to get these dogs into the show rings and Eva did. Robert Zoiler declares in his writings that "I'm convinced a lot of mistakes were made. I've seen a number of dogs claim to be Malamutes that weren't even close…" (Riddle, p.57) I believe he felt a bit guilty as within the A.K.C show rings these Alaskan Malamutes were shown with blue eyes. Some of the dogs shown in the Baltimore show of 1947 resembled Saint Bernard's! Eva had gotten a standard for the breed written up as the American Kennel Club had required this. The standards did not say that the Alaskan Malamute had blue eyes, but that they should be brown. (I do believe Eva really liked those blue eyes though and somehow she got the Siberian Husky registered into the American Kennel Club also).

I went back into the archives and made some sketches of the sketches that were made by the naturalists who were around during this period in time. These sketches are of the dogs that were sent to Alaska on ships from the states to help in the gold rush.

Most of you doggy folks out there have read the dog stories of the gold rush. You know that Saint Bernard and German Shepherds were sent to Alaska as well as hundreds of mutts. The gold rushers needed weight pulling animals so dogs that could do the job were sent up to Alaska on ships. The Eskimo Indians didn't require weight-pulling dogs for their survival the gold miners did! What was more important to the Eskimo was that the dog could pull the family sledge long distances in very cold climates with little or nothing to eat. They most certainly didn't care about what society folks cared about like having pedigrees or making sure the dogs bred true!

The Indians never used big dogs, those dogs ate too much and the Eskimos couldn't afford that. Seems to me that the Indians would rather prefer to eat the larger dogs. Maybe that's why there weren't any large beefy dogs up there!

Eva Seely got up to Alaska long after the many different types of dogs got there so maybe Eva really believed the dogs she saw up there to be the dogs of the Malamute Indians. Who knows? The fact is that the dogs she brought back were mutts of her choosing and that makes the Alaskan Malamute less than 100 years old.

All this is written in the Alaskan Malamute books. What I say is no secret and I'm not complaining but please, don't tell the world that this is the oldest of all breeds of dogs from the North. Nor that these dogs were the "natural" true huskies of the Northern World. That kind of information believed by the public really messes it up for the real Northern dogs that were there ahead of the Alaskan Malamutes.

Here are the words taken from the A.K.C breed standards: "The head is broad and powerful as compared with other *natural* breeds." That statement alone tells me that the Alaskan Malamute is supposed to be a natural breed of the Northern World. Maybe even around for thousands of centuries! Maybe I read it wrong, if I did forgive me, but that's how I read it.

I'm thinking that what I have to say might come as a shock to some of the readers of this book. Especially to the proud Alaskan Malamute owners who have believed all their lives that the Alaskan Malamute was the real true Northern sledge-pulling dog. I suppose many are saying, how dare she and where does she come across with this? I have done the research. I have used my common sense and years of experience in my life to put it all together.

Now don't get me wrong, I love the old fashion Alaskan Malamute! A beautiful dog this dog turned into and with a great personality as long as he doesn't see a cat! The Alaskan Malamute has never been out in the public limelight so that breeders could destroy the structure of this magnificent animal. How is that you say? Well, you see if the public purchases your particular breed of dog and your breed is in high demand, then a lot of breeders would breed that dog to have puppies that would sell at high prices. It's a business, a way to make money. After all, this is the land of the free, isn't it? The problem with that is that some breeders know nothing about selective breeding or what breed standards are. Or maybe they do, but don't care? Well, that is one of the ways that a breed deteriorates in health and temperament, when folks do not selectively breed to a dog that would complement the bitch. Then you have owners who don't even see the faults in their dogs. Maybe they don't really know what type of temperament their little dog is suppose to possess. Maybe they don't care. "Gee, its only one litter, what is that going to hurt?"

Well, it's the person that buys your pup and then breeds it. That is where those bad genes may pop up or be hidden. Will the backyard breeder know this and what will they do about it? Pretty soon the whole barrel of apples is rotten. Gee, I remember when the American Cocker was the best little dog in the world, it was so happy, wouldn't bite anyone. Today one has to approach a cocker with caution.

Most popular breed —- Every year the American Kennel Club puts out its list of the most popular breed of dog in America. If you really look into that you will find that the most popular breed may not be the most popular breed, rather the breed that has had the most puppies registered. That's how the count goes. What is wrong with that you might ask? Well if my dogs have 18 puppies in a litter and the little Chihuahua only has one pup per litter which dog would be declared the most popular breed?

Anyway, I digress again. Back to the Malamutes…

During the beginning of the creation of the Alsatian Shepalutes I had to find the perfect Malamutes that would bring that wonderful and loving Malamute temperament into my lines. As I tried to find the right one I began to see the more thin boned Malamutes with a hyper ness about them and a high pitched bark! Heck, there's no way I would breed that into the lines. So I concentrated on breeding only with the old strains; Malamutes with large heads and thick bodies, those that were not hyper active.

Malamutes were chosen to be bred with German Shepherd Dogs that passed certain tests of what the public wanted. In my opinion, the Alaskan Malamutes body structure (according to the Alaskan Malamute Club standards) cannot be improved upon. All dogs should have such a body!

Other Northern Breeds —- Now, there are many registry clubs like the American Kennel Club who have breeds of dogs registered within their club that prove to be the oldest of the Northern breeds! The American people need to understand that A.K.C is not the truth of all breeds and that the world is bigger than just America. The following are descriptions of other Northern dogs that have been categorized, named and registered as the true Northern breed! The reason I must include these other Northern breeds in an Alsatian Shepalute book is that the Alaskan Malamute, in my opinion, carries these breeds all together in its background. I also believe that it doesn't hurt to educate yourself to overfull. Knowledge stimulates the mind.

I cannot provide you with a full description of all the Northern breeds. (Too much typing). You can look them up on the Internet. But here is my take on a few of them. Put them all together and you have the Northern dogs themselves. Sledge hauling dogs after the influx of the many different breeds that went up to Alaska to help in the gold mining expeditions. So there just isn't the real "Kingmik" any longer and that's why no one can find it. Too many dogs interbred with it.

Eskimo dog —- This dog is known to the Inuit people as the "Kingmik" or "Qimmiq" or ("dog" in the English language). The Eskimo dog has also been called a Greenland dog, a Gronlandshund and an Inuit Husky. Probably because the name "Eskimo dog" was a slang name for all dogs belonging to the Eskimo people.

Eskimo Dog

This dog is said to come from Greenland. According to A Standard Guide to Purebred Dogs, "The Eskimo Dog originated in Greenland and is the most generally useful of the hauling dogs of all the Arctic countries" (Glover, pg. 425) Canada claimed it as their own and when somebody asked "Say, what kind of dog is that?" Somebody else answered "why; he is an Eskimo dog. You know, that dog that the Eskimo's used." And of course this dog is the dog claimed to be the only dog of the early arctic peoples. Being such, this dog guarded the family and hauled supplies on their sledges. He was accepted by Canada as "the" Eskimo dog in about 1900. Hmm? Wasn't that after the influx of dogs from other parts of the world? Anyway, it has its own pedigree and it is registered with the Canadian Kennel Club (C.K.C). He is a big dog. Stands about twenty-five inches at the shoulder when measured from the shoulder to the ground. Hmm, a lot of meat on this dog, don't you think?

The Yukon dogs —- This is the true Eskimo dog. (He-he) These Eskimo dogs were known by the miners as Malamute. That word "Malamute" described a tribe of Eskimo's with that name who lived at the mouth of the Yukon. A bit confusing? Anyway, this Eskimo dog stood about as high as the Scotch Collie, which it also resembled. Let me repeat that: this dog resembled a Scotch Collie! What do you think? Does the Alaskan Malamute look to you like a Scotch Collie?

This dog had a thick short neck a sharp muzzle and oblique eyes. It also had short pointed erect ears and a dense coarse coat of hair. This

Eskimo dog also had a bush tail that it carried tightly curled up over its back and came in a variety of colors. This Eskimo dog that the miners called "Malamute" ranged in color from a dirty white, black and white and black. There was also another color called a grizzled gray. To some that grizzled gray made this dog appear to be part wolf.

Sivash dogs —- These dogs were not as large as the Yukon or malamute dogs. Mostly the Indian tribes of the interior owned the Sivash dogs. These little dogs were tougher than the Yukon dogs.

Greenland Dog

Greenland dogs (Gronlandshund) —- Say, didn't we just read that the Eskimo dog came from Greenland? Maybe this dog is the father of the Eskimo dog? To me this dog looks like the closest to what I would think an Indian's hauling dog would look like, yet it still seems to me that even this dog has the resemblance of some of the dogs brought over into the Northern lands from those ship sketches.

A Standard Guide to the Pure-bred Dogs, written by Harry Glover, on the Greenland dog starts out with, "Considerable difference of opinion still exists concerning the names of the arctic sledge-hauling dogs" (p. 427) Hehe, I guess so. "The Federation Cynologique International" registered this breed. What do you mean you never heard of them?

This dog is a great watchdog, as he doesn't like strangers much.

Leonberger

Leonberger — It is amusing that in 1907 the Leonberger was dismissed as a cross between a Newfoundland and a Saint Bernard when it was first introduced into Britain. I really do find that funny, but at least somebody was paying attention!

One writer of the time stated that its merits were only recognized by the enterprising gentleman who had presented it as a 'new breed' (sounds like Eva to me) I wonder if the dates are the same?

Iceland Dog

Iceland dog (can you see the outline of a mask?) —- This is a small dog, twelve to sixteen inches at the shoulder. Folks say this dog comes from Iceland. The Iceland dog does not have much in the hunting instincts, but is a great little herder. It is said that the Iceland dog can trace his history back to the Vikings. That is around 874! He is a member of the Spitz group. He weighs about twenty to thirty pounds. He was known in Britain and was recognized by the British Kennel Club (B.K.C). The Danish authorities accepted a standard for this breed in 1898. He is similar to a small elkhound with less coat. He is what is called a spitz type. The spitz type of dog was probably a lot of dog's grandfathers.

Lappinporokoira

Lappinporokoira — There was another spitz dog called the Lapland Spitz. (Maybe that same grandfather, Hehe) That dog was bred in Southern Finland to the sheep herding dogs of Germany (German Shepherd Dogs?). It was also bred to the collie of that area. Maybe that Scotch Collie that we heard about that was taken to Alaska? This dog was approved as a separate breed by the F.C.I in 1946 and resides in Scandinavia.

The Alaskan Husky —- While I was growing up, any dog that resembled the Husky or the Malamute was called an "Alaskan Husky dog" (we didn't know the registered politically correct names, you know). We'd say "look at that Husky" or "there's one of those dogs from Alaska." There was no such breed! The name suggested a group of dogs that pulled sledges and came from or lived in Alaska. Simply that. Well, now guess what? Some folks have decided that they are going to find such a dog. Yes, that must be the REAL true Alaskan Husky!

In Conclusion — By the way, I would like to note that many people thought these dogs were bred by wolves as many folks associate the wolf gray coloring of any dog to be the result of a wolf cross! The early naturalists that went to Alaska to record their findings continually searched for anyone who had true knowledge of dogs breeding with wolves. After many interviews with the natives, it was written by these old timers that no one could find any Indian or Eskimo that would declare that it was indeed a truth that a dog had bred with a wild wolf. For further information concerning wolf/dog breeding you may like to read "Of Wolves and Men" by Barry Holstein Lopez.

German Shepherd Dog

At the beginning of the German Shepherd Dog's creation there were 3 distinct types of sheep herding dogs in the vicinity where a man named Capt. Max Von Stephanitz lived. These 3 types of shepherds were:

1. Herding dogs from the mountains near the Swiss Alps.
2. Sheep herding dogs of the valleys or central Germany.
3. Herding dogs from the rocky hillsides of Germany and/or France.

Sketch of early German shepherd dog

Sketch of the Central Germany Sheep herding dog.

Southern Sheep Herding dogs.

As you can see, these dogs varied greatly in size, coat, tail carriage, ear set and even the way they herded.

Capt. Max brought them all together into his own breeding program. He selectively breed these dogs by hand, choosing only those pups that showed the characters of a great working and herding dog, one that would fit his breed standards. He did not care about the beauty of the dog nor did he care about the politics. He did put these new dogs through his own testing programs for sound character.

In 1891, a group of persons interested in these dogs formed a group called the Phylax Society, which supported this breed. This group of folks was primarily interested in the conformation aspect of the dog. There was not a great movement of the society and it withered away.

The Captain then formed the "Verien fur Deutsche Schaferhunde" or the SV as it is called even today. This group of folks adopted a working standard and that standard still presides over this breed today in the Schutzhund sport along with the proper German registration papers.

The German Shepherd Dog —- The first owners of the German Shepherds in America entered their sheepdogs with the American Kennel Club (which was already in existence) in the miscellaneous class, as there were no German Shepherd Dogs registered. That was in 1906. In 1908, that same dog (Mira) was exhibited in the miscellaneous class in New York. This time, she was registered as a German **Sheepdog**. In 1907, Mira was classified as a **Belgian Sheepdog**! The Belgian sheepdogs were combined with the **German sheepdogs** during that time. Both were shown in the miscellaneous class.

In 1913, two fanciers of the German Shepherd Dogs appeared here in the states and started up the "German Shepherd Dog Club of America".

The judging of this breed in the A.K.C rings was poor as nobody really knew the standards of the breed so square; large dogs were continually chosen to win. The official name of the breed was the "**German Sheepdog**" even though the club was called the German Shepherd Dog Club of America. The A.K.C then changed the name (as they will) to the "**Shepherd Dog**." (That's when they should have called it the **American Shepherd dog**!)

In 1931, the name changed back to the German Shepherd Dog. (Nobody was very fond of the Germans at that time and all mention of the German name left a bad taste in folk's mouths.) In Britain, the breed was known as the "**Alsatian Sheep Herding Dog**." In 1979, Britain decided to go along with the rest of the world and call the dog by its proper name the "**German Shepherd Dog**."

Today in the American show ring, judges and breeders of the German Shepherd Dog strive for that long reach which makes this dog appear to be prancing or pulling the handler around the ring in the high stepping fashion of the German army.

Soon three distinct types of German Shepherd Dogs began emerging throughout America.

1. The American Show Shepherd
2. The Schutzhund Shepherd
3. The Backyard Shepherd

The American Show Shepherd —- The standards of any breed are supposed to protect the type and character as well as the physical body structure of a breed, yet the American German Shepherd dog breeders, along with the American Kennel Club, don't adhere to the standards of Germany. They made up their own standards for the German Shepherd

Dogs of America a long time ago. The show breeders and the A.K.C judges are proud of the fact that the German Shepherd Dog covers more ground per step than any other breed of dog. Folks with money can get these high-priced show dogs which aid in dividing this breed further and the public gets the so called "backyard" shepherds.

German standards differ from American standards in that the German standards call for a 50/50 chest verses leg length. The American standard calls for a longer leg, therefore, a shorter chest. That in itself separates the two dogs as it once separated the English from the American Cocker Spaniel. Let me repeat that. The German standards and the American standards are not the same! How come? Do we think that we can just change the standards of a breed to fit whatever we want? What is that called? Creating your own breed?

The Schutzhund German Shepherd —- Now let's discuss the German Shepherd Dogs from Germany or the Schutzhund dogs. Most of the Schutzhund German Shepherd Dog owners are upset about the American German Shepherd show dogs and they continually talk them down. In their opinions, the American German Shepherd Dog of the show ring does not follow the standards of Germany where the dog originated. The American German Shepherd Dog Club formed their own standards of the breed when they hooked up with the American Kennel Club. Those changes in standards alone divide the breed.

In many Schutzhund breeders' opinions, the American German Shepherd Dogs lack drive and the natural instincts that the Schutzhund shepherds possess. They would never breed an American Shepherd Dog with any of the German Shepherd Schutzhund Dogs. As a witness to all three types of German Shepherd Dogs within America, I tend to agree. That does not mean that I do not love and admire the American German Shepherd show dogs. As far as I am concerned, the American show Shepherd's have a better personality as a family companion dog than the Schutzhund Shepherds have. That's the problem; you see the German Shepherd Dog was bred to be a working dog.

The Schutzhund shepherds are the correct working German Shepherd Dogs. I personally would never again in my lifetime own a Schutzhund German Shepherd Dog. They are far too energetic for me. But for those that love the true German Shepherd working dog there is no other breed! The real German Shepherd Dog is the working dog!

Schutzhund is both a sport and a means for preserving the working character of this breed. Impartiality tests evaluate the temperament of the dogs. A Schutzhund judge tests and scores dogs in three phases. Phase "A" tests a dog in the area of tracking, phase "B" tests in the area of obedience and phase "C" in protection. All of these scores and tests are one heck of a way to evaluate the true working dog!

The Backyard German Shepherd —- Now, enter the backyard American German Shepherd Dog. This must then be categorized as the "mutt" of the German Shepherd Dogs, as it possesses both traits (German and American) and breeds indiscriminately within the two. These are the dogs the American public ends up with. Though they are registered German Shepherd Dogs I have found that their pedigrees are filled with the German Schutzhund working dog lines.

The Schutzhund dog is a highly motivated type animal with very strong drives and most normal everyday folks cannot handle this type of shepherd. Many German Shepherd Dog owners came to me asking me to help them so I know this for a fact.

Schutzhund German Shepherds were bred to work in the Schutzhund competitions. Schutzhund competition is the correct show ring, so to speak, for the correct German Shepherd working dog. This sport brings the breeding importance of the high drive Schutzhund German Shepherd Dogs to the area of the mouth or natural biting instincts. A hard bite is necessary in their competitions. Schutzhund German Shepherd Dog puppies are selected for breeding and training if they have a hard bite and attack a rag with enthusiasm and aggression.

There are two great divisions within the German Shepherd Dog lines. To me that is a problem and it is one of the many reasons for the start of the new breed, the Alsatian Shepalute. I know for a fact that the average homeowners that have bought the Alsatian Shepalute have found in them what they could not find in the German Shepherd. A few German Shepherd breeders have asked me why I didn't just breed the German Shepherd if I liked them so much. They suggested that I might work on perfecting the breed if I thought it needed it. I think the readers can see that wouldn't work. The German Shepherd Dog's standards call for a working dog and the Schutzhund German Shepherd Dogs fit those standards. The A.K.C doesn't come close to the German show rings, which are in fact working trial rings, so to speak, and in my opinion the only way to judge for the true German Shepherd Dog. I did not want a working dog but a companion dog. Therefore, I refused to breed the German Shepherd working dogs and turn him into a companion dog. I did not feel it was right then and I still don't.

There is a place in this world for all the breeds, though, the short-haired American cockers and the cotton-coated American Cocker, the American German Shepherd and the Schutzhund German Shepherd. I say, just make up a new standard and stop lying about your breed. Be proud of that drive and biting instinct and be proud of that high stepping long reach and that gentle show dog. Come on; change the standards to fit the dog you are breeding and showing then the American public would not be confused. If you want the German Shepherd to be a great companion dog then change the classification and the name. If you want an American Shepherd and you don't want to fully accept the German standards for their own breed, and then change the name.

That's how new breeds are formed and if you take a look at all the new breeds being formed everyday for other reasons, perhaps one would consider a good reason to split a single breed into two different breeds. It helps protect the desired traits of the cotton coat verses the short field coat, or the working (hard biting) Schutzhund Shepherd verses the mild mannered American show Shepherd.

Sketch of Diamonds Willow of Cold Springs

English Mastiff

I will not go into all the official standards of the Mastiff but will only give the reader the qualities that I felt would benefit my new breed. These qualities were selected from pups out of many generations. If you wish to do your own further research into the Mastiff, look for books that contain the standards of this breed along with a point system for which the English Mastiff is judged upon. Competition strives for points thus breeders will breed according to how many points they will receive in order to win in the ring. That is the dog show business and that's the way it's done.

General character—— The Mastiffs that I have come into contact with during my lifetime have been skittish, shy animals. The males have been known to bite strangers. Some Mastiffs have rushed and growled at me when I approached their property. The females that I have known are fear biters or simply shy and will run away with their tails between their legs and their heads lowered. Some Mastiff owners and breeders will of course be offended. Maybe I wasn't around the right group of breeders? All I can say is how many breeders do I have to be around to see that shy Mastiffs are all over the place? I know you can modify the behavior and with enough strangers approaching your dogs you can get them to expect no harm. The basic dog is still shy no matter how you train him. I've seen Mastiffs in the attack ring also. Yep, is that bravery or fear? I'm not going to debate the English Mastiff I am only reporting what I have seen and that is what I saw. So why breed to them you ask? Well, let's go on and see.

Mastiffs do not like to be lifted off their feet. I found it difficult to take my Mastiffs for rides in the car. Training them to get use to walking on different types of objects was also difficult. After owning and training the German Shepherd and Malamutes and then the Shepalutes, It was hard to own any other breeds that did not train or respond so easily. Mastiffs also do not care for strangers. Within the family circle of friends or in the grooming shop (familiar territory) the Mastiffs did ok. Once this breed got to know a stranger it was very loving and child-like in nature with a clownish personality. They love to be close up on your body, even leaning their massive bodies into yours so that they may feel the security of the owner. Maybe they will just put one of their heavy paws on your foot just to know you are there while they nod off to slumber.

General description of the head of an English Mastiff —- In the American Kennel Club show ring, being judged as a representative of the breed, a Mastiff must have a large massive head. It must be very wide. The skull should be somewhat rounded between the ears with the forehead slightly curved and with lots of wrinkles, which are particularly distinctive when at attention.

The face is short and deep, as is the muzzle, deep from the nose to under the chin. The eyebrows are slightly raised giving the Mastiff an inquisitive look. The eyes are brown and the ears are small and v-shaped being rounded at the tips and lying close to the head. The ears should be dark in color, as with the muzzle. The teeth should come together in a scissor bite but an undershot jaw is permissible providing that the teeth are not visible when the mouth is closed. The nose is always black and the neck is very muscular without any loose skin. The chest is wide and deep with the forelegs and feet being straight, strong and wide apart with heavy bones. The color of the Mastiff should be apricot, silver fawn or dark fawn-brindle. The male dog should be 30 inches from the ground to the shoulder with the females being a bit smaller.

The following is the point system that the A.K.C judges are suppose to use as a guide in their show rings. This point system enables them to award prizes and ribbons to the dogs that best represent the standards of the breed. The Mastiff, whose point's total more than the other Mastiff's it is competing with, will be awarded a trophy or ribbon. They will get a lot of recognition and the stud fees can increase substantially. Let's take a look at what a breeder would breed for and/or what a judge would look for:

Scale of points:

Face and muzzle	12
Skull	10
Ears	5
Eyes	5
Total points for the head	32
General character and symmetry	10
Height and substance	10
Back loins and flank	10
Chest and ribs	10
Total points for the body	40
Forelegs and feet	10
Hind legs and feet	10
Total points for the legs	20
Coat and color	5
Tail	3
Total points always equal	100

Studying this tells me what the Mastiff (as a breed) is strong in. It also tells me what the breeder's breed for and which character will be strong in the dogs I breed with. (If they fit the standard of the breed). Looking at the chart, one can see that a breeder would breed for the face and muzzle. On the scale of points, one of the most important features of the English Mastiff is his face and muzzle. If you go into a ring with a great face and muzzle on your Mastiff you will get 12 points! The skull will also give

34

you 10 more points and if you have the face and muzzle, I am sure you'd have the skull also. The eyes and ears are easy so the total head points all together for the head would give me 32 points. Remember, this is how I would breed if I only cared about winning trophies, recognition, breeding rights, stud service fees and puppy sales. (Monetary value)

Now, what would you focus on if you were a breeder? Well, that's what I wanted, a very large dominant head.

The next to be focusing on would be the Mastiff's total body. The points for the total body are 40. So you see judging by this scale of points, breeders would breed for these two characteristics; total body and head. That was exactly what I was looking for. Those qualities and the Alaskan Malamutes body points give these Alsatian Shepalutes a great start. One could just imagine the width on this new breed!

Let's take a look at what I would not care too much about. That would have to be the tail at 3 points and the coat and color at 5 points. The most important physical characteristic of this mammoth breed would not be such a big deal in this point system. All the feet, forelegs and hind legs, which must work properly to keep this huge dog up on all fours, only counts for a total sum of 20 points! No wonder I saw so many deformed feet in the English Mastiff as a breed.

I also have to read the pedigree to find out what genes will be passed down into the new pups. Now, here may be a problem. Not all pedigrees are the true ancestor's of the pup I am purchasing. If the world was perfect and nobody lied or falsified pedigrees then I could believe a pedigree. As it is, I do not believe what I haven't seen or know for a fact to be true. I will take it for what it is and I will keep in mind that it might not be the right pedigree. So, I pick a pup from what I see. And, of course, I will check out all the dogs in the kennel and I will do my own little testing of each pup.

Since I was looking for more body mass and a larger head to be bred into my pups, while also keeping with the good nature of my dogs, I chose a Mastiff pup to complete the look of the new breed. (I didn't want a lot of Mastiff in the lines as I found these dogs to have bad feet and to be skittish or fear biters.)

The Mastiff I chose to breed with came from obedience title lines. This told me that the character of the pup off this side of the lineage would be able to learn and follow commands. It also tells me that this side of the gene pool probably could not compete in the show ring. Or maybe the owners didn't care to show their dogs in the show rings.

My puppy Mastiff had a show sire that gained his champion title at the age of 18 months. This tells me that his genes were great in the body and symmetry field along with the head and skull. Just what I was looking for! Out of the entire litter, I chose the pup with the largest girth or body mass. I did show this puppy in puppy classes and the A.K.C rings until a judge miss-handled my dog and the dog growled at her. (One would think a judge would know how to approach the breed she was judging). Even the crowd was displeased and spoke to me about it when I got out of the ring. Some of the folks were paying attention!

My Mastiff pup also had a slight undershot jaw that was not noticeable until a judge pulled at it so that it got hung up on her teeth. That was at another show where the dogs were showing in the grass. Everyone knows that Mastiffs with poor (turned in) feet were always shown in the grass where the judge wouldn't notice! And this judge didn't. I would rather breed my Mastiff to a dog with good feet than to care about a slightly undershot jaw! Feet are very important especially when you are a dog that weighs 215 pounds!

The author and one of her mastiffs at an A.K.C show

My Mastiffs gait was unsurpassable and her body structure was more massive than the other Great Dane looking Mastiffs that had been showing in the California rings, which means she was shorter.

I did not envision a Mastiff for the Alsatian Shepalute but I did need the bone structure and body mass to reach the intended goal. It took several generations of selective breeding to properly dilute the genes I didn't want and to keep the Shepalute's high intelligence without that dominant shyness. Shyness is a dominant gene and will continue to show up in all pups until properly eradicated. (That means completely gone.) It is one of the hardest characters to eliminate! A good breeder must remember that. Don't ever breed a shy pup!

Saba. Second-generation (F2) Alsatian x Mastiff mix bred back to an Alsatian Shepalute. Five-months old.

Northern wolf

The Wolf
(All that we know about the wolf is mostly imagined)

I am including a bit on the wolves because the standard of the Alsatian Shepalute calls for this breed to resemble the wolf. Giving the reader some information on wolves will enable them to picture what the true Alsatian Shepalute should look like. I would like to inform the readers that I am not a wolf expert. I have only studied on the subject and have not gone out in the field for observation of the canis lupus (wild wolves). It's too cold up there for me. (Hehe) Of course, I do not believe everything I read or hear. Human beings write what they wish and no human is perfect. The following are my opinions and are meant to make the reader think, "What do we really know about the wolf?"

Grouping of wolves: I do believe folks have tried to invent more than what is necessary in the different species and subspecies to be noted. Somewhat like what happened to the Eskimo dog.

The following are supposed to be species of wolves: (When I went to school I learned that different species were unable to breed or produce viable or breed able offspring, if any).

Gray Wolf, Timber Wolf, Lobo, Asian Wolf, European Wolf (canis lupus)

From the species comes the subspecies:
Siberian wolf: (Canis lupus subspecies)
Alaskan arctic wolf: (Canis lupus pambasileus)
Mackenzie valley wolf: (Canis lupus occidentalis)
Alaskan tundra wolf: (Canis lupus tundarum)
(Canis lupus alces)
(Canis lupus hudsonicus)
Eastern wolf: (canis lupus lycaon)
Buffalo wolf: (South Dakota) extinct
Custer wolf: (South Dakota) extinct
Dire wolf: (c. Dirus) extinct
Texas red wolf: (Canis Rufus) (notice it doesn't say canis lupus Rufus)

Descriptions:

Siberian wolf—- It was told to me that these were the largest of the wolves. (Is that like the Eskimo dogs?)

It is my personal believes that these are the ancestors of the arctic and dire wolves before the Pleistocene age, which traveled down into the Americas.

Arctic wolves (Canis lupus pambasileus) ——The largest wolves have been found along the arctic landmasses. Young says that Ernest Thompson Seton killed a wolf at Aylmer Lake which being very gaunt weighed 88 lbs after being bled. This wolf would probably weigh about 100 lbs in good condition. Seton, a naturalist, said that dogs were these wolves' favorite food sources. (Riddle, p.66) Others have observed the same. So, are these the largest wolves or have the largest wolves all but disappeared?

The Canadian wildlife service lists only three species of arctic wolves.
1. Mackenzie Valley Wolf, (canis lupus occidentalis)——- I believe this is the same wolf as the Arctic Wolf and sub categorized as the Mackenzie Valley Wolf only because it lived in this area. Of course, they saw this wolf in the Mackenzie valley, thus the name.

2. Alaskan Tundra Wolf (Canis lupus tundarum) —- a name proposed by miller in 1912. Pure white on cream white with darker hairs on the back and tail. Maybe just a different color of the Artic Wolf?

3. Hudson Bay Wolf (canis lupus hudsonicus) —— another Arctic wolf named the Hudson Bay Wolf because of its existence in the Hudson Bay.

North American wolves: (Canis lupus) gray wolf or timber wolf —— carnivore of the family canidae. (Canis lupus). An inhabitant of both open and timbered areas, the gray wolf was once found throughout North America and Eurasia. It has been eliminated from much of its range and is now found primarily in Asia and in North America from Alaska to the Northern plain states. It carries its tail high when running. It is 6.6 feet long including its tail (20 inches) and weighs about 100-110 lbs. These wolves live in packs of several to two dozen.

Eastern Wolf —- (Canis lupus lycaon)

Buffalo Wolf (South Dakota) —- I do believe this wolf was the direct relative of the Dire Wolf. The largest known wolf. The Buffalo Wolf that one looks up today is not the Buffalo Wolf that I am writing about in this paragraph. In my opinion, this wolf was large and looked similar to a buffalo with large shoulders with a broad chest, (c. Dirus) common in Western North America during Pleistocene. This wolf was half again as large as the modern gray wolf.

The Buffalo Wolf is now extinct and another wolf has probably moved into South Dakota. Wonder what they call that wolf?

41

Custer Wolf—- (South Dakota) a relative of the Buffalo Wolf, now extinct.

Texas Red Wolf (canis Rufus) —- formally called (c. Niger) endangered. This wolf is the smallest wolf in the world. It is longer legged in proportion to height than the gray wolf. It has a more massive head and a broader muzzle than the coyote. It also has a larger foot. The color of its coat is a yellowish brown to reddish with a cinnamon color on the muzzle and around the eyes. It weighs roughly 30-35 pounds and is 19-22 inches at the shoulders. It is said to be a true subspecies. I believe that the Red Wolf breeds with the coyote (another species); therefore, I do not classify this canis Rufus as a species of wolf.

The Texas Red Wolf has a black phase which, once found in Florida, was sometimes called the Florida Black Wolf. This so-called black wolf is supposed to be the same as a Texas Red Wolf only black. So its head and muzzle is broader and longer than a coyote, but, as I stated, it is said to breed with coyotes. A wolf will kill a coyote. In my opinion, this so named Texas Red Wolf is not a wolf at all.

Texas Red Wolf.

General Characteristics

Size—- Stanley P. Young is said to have shot the largest wolf ever killed. It is claimed that it weighed 175 pounds and measured 38 inches tall at the shoulder. (Lopez, p.18) (It was never proven)

While these figures may be correct, most reports of gigantic animals are lies or truths unable to be substantiated as with Stanley Young's description.

In the far North, male wolves average 95 to 100 lbs. Young said that he killed a wolf in the far North that weighed 116 lbs. That is recorded as being the maximum. Northern females range between 65 and 100 lbs.

In studies of Isle Royale wolves in Ontario, Canada, Rutter and Pimlott give average sizes of 61 lbs and 80 lbs maximum for the males. One female weighed only 39 lbs. (Lopez, p.19)

Breeding—- Wolves can reach sexual maturity as early as eight months. Being young they are not allowed to breed by the dominant dogs of the pack. Dominant females harass young females (though I personally believe that sometimes a young female may become pregnant). The now pregnant young bitch instinctively backs out of the pack to whelp on her own (or she may get harassed so much that she leaves the pack for her life). Then again a pregnant bitch does turn into a raving whirlwind of clashing teeth! This pregnant bitch may just take down the dominant bitch and a trading place of hierarchy would now take place.

If a young female gives birth to pups while in the pack, her pups will be eaten, or if the dominate bitch has a litter that dominate wolf might take the pups for herself to wean. Some wolf researcher might count all the pups (maybe 12 or 14) as belonging to the dominant female and would note this in the records that the other females did not whelp pups. The dominant female may also eat her rival's pups if she determines she does not like the smell of the new pups. She may also eat her own pups if overly stressed or because of a problem with other members of the pack or because of a fear ingrained in her instinctually.

Litters are generally three to five pups. Larger litters may be the result of younger female's pups being stolen by the dominant bitch.

All members of the pack will dig dens. Wolves have no natural predators and roam up to 50 to 100 miles a day. Food will be regurgitated for the pups if the pups lick the muzzles of any pack member, or food may be brought to the pups.

Character—- Wild wolves have survived because of the flight instinct. A natural instinct to flee, a genetic timidity, a nervousness or restlessness. Therefore, wolves pace. They have a high energy level because of this nervousness and a very quick responsive defense system. This is a dominant trait, always.

If it were a true fact that wolves only have 1 heat cycle per year, then that would put the wolf in a different species category than the dog or canine domesticata familiar. That is an extremely big difference between the two. That would indicate to me that the wolves and dogs are even more widely separated genetically than I thought.

Pack members are usually related. How could they not be?

Sometimes there is a pup that is constantly picked on by other pack members and is driven away or killed. Old dominant wolves along with their mates may be driven from the pack forming new packs or they also may be killed. Wolf packs usually number five to eight without too much conflict and can grow up to thirty-five individuals if one considers it a pack when they do not stay together. I do not.

Wolves run after animals that flee. Pups chase leaves and other objects when those objects dart or move away quickly in front of them. This is the inbred genetic behavior that I call "the prey-orientated behavior."

Wolves kill dogs—-Wolves kill dogs, coyotes, and even other wolves that are submissive, sick or injured.

Capt. Lyons reports, "The passion which wolves showed in killing dogs made this (wolf/dog cross) impossible. Wolves… invariably kill dogs." (Riddle, p.66) (At that time in history they hadn't heard of artificial insemination.)

I hope that this chapter on wolves will give you a little insight as to how the Alsatian Shepalute should look. Of course, all the bad characteristics of the wolves are undesirable traits. Concentrate on the look, the stare and the secrecy of the wild animal and incorporate that into a domestic, calm, loving animal and you would have the Alsatian Shepalute!

Chapter 2

Genetics

This page is not intended as an educational text on the complete theory on genetics of the carnivore's mammalian, if one wants to learn more on this subject I shall suggest the following books in the order of my favorites listed first: Refer to bibliography.

How to Breed Dogs by Leon F. Whitney, D.V.M.
The New Art of Breeding Better Dogs by Kyle Onstott
A Standard Guide to Purebred Dogs compiled and edited by Harry Glover
The Inheritance of Coat Color in Dogs by Clarence c. Little, sc.d.
General Zoology by Gordon Alexander
Of Wolves and Men by Barry Holstein Lopez
Canines and Coyotes by Leon v. Almarall
Encyclopedia Britannica (all information concerning genetics)
The Wild Dogs in Life and Legend by Maxwell Riddle
The Complete Alaskan Malamute by Maxwell Riddle and Eva b. Seeley
German Shepherd Dogs by Francis Kern
The Encyclopedia of Dog Breeds by Ernest H. Hart
German Shepherd Dog by Jane Bennett
Rare breed handbooks
All old and out-dated dog books
(And any other books you can get your hands on. Education is the key to understanding.)

Simply put the Alsatian Shepalutes are the offspring of Selected German Shepherds that have been bred to superb Malamutes with an influx of the large boned English Mastiff.

First Generation Hoss of Cold Springs. Mal x Shep.

Mutts and Pure-breeds

What is a mutt? ———- A mutt is a dog whose lineage, family tree or ancestors are unknown.

Even if the dog looks like a purebred dog it cannot be identified as a purebred dog without a written lineage and tattoo. A mutt is also known as a "Hines 57." (Don't go looking for the breed Hines 57)! What that means is that the dog in question has so many different generations of different breeds in its family tree that this dog is a pure mutt.

There are dogs that have two different purebred dog parents. These would be called F-1 generations, or first generations. They may be called cockerpoo, terrypoo, shitzapoo, maltipoo, and many other forms of the names of two purebred registered dogs. These dogs have been given registrations from Mutt Clubs of America, or the Poo Club and whatever other clubs are out there. First generations do not breed true; therefore, first generational mixes are mutts also. A pure of breed (purebred) breeds true.

A hybrid is an offspring that was produced by a male and female of different species. Example: Mule. Different breeds are not different species. They are the same species carnivore domesticata or domesticated meat-eater. That to me does not include the wolf. If a wolf were a different species than the domesticated dog then a dog x wolf (or a pup from a mating of a dog and wolf) would be a hybrid. A pup as a result of a boxer and Rottwieler breeding would not be a hybrid. Alsatian Shepalutes are not hybrids.

What is a pure-breed animal? —— A pure breed animal is an animal that consistently reproduces itself. This means, the same height, width, size, weight, eye position, ear position, tail carriage, trot, gallop or pace with the same coat consistency and the same skull dimensions.

A purebred dog is a canine that has been bred with various selectively chosen other dogs that possessed some trait or traits that the creator or breeder of that strain of dogs wished it to obtain. When the offspring are bred to the same, they too will produce like progeny.

Create: 1. To cause to come into existence; originate. 2. To be the cause of; occasion: To create interest. 3. To produce. (Funk and Wagnall's Standard Desk Dictionary, p.150)

When the standards of a breed, any breed, are written down and followed then the formation of that animal comes together. The bonds of the genetic code are more closely related and breeding to only these dogs, which resemble that standard, would solidify the progeny. How many generations it takes to breed the dogs so that all puppies would resemble all other puppies, dams and sires, depends on how long it takes to get the results described in the written standards.

For example, say you wanted to create a new breed of dog, a dog that had a drop ear instead of an erect ear, yet with all other parts of the body being exactly the same. Well, when you finally have all pups in all generations with drop ears and they continued to produce only drop ears, then you would have accomplished your feat. (Erect ears are dominant. Ear positioning varies. I call it a sliding scale with the erect ear pulling stronger than the drop ear. Therefore, erect ears are the dominant of the many different ear positions).

A truly magnificent correct American Cocker! Not one of mine.

Another example of creating a new breed of dog would be the American Cocker Spaniel. The early American cockers had flat coats that did not inhibit the job for which the cocker was created for in the first place. The dog in this picture is considered by me as still having too long of a coat. But it would not hinder him in the field.

No where in the standards of the American Cocker Spaniel does it say to clip the dog's coat. The American Cocker Spaniel's standards of the breed calls for a flat coat that does not hide the cockers true lines and one that is easy to care for. The standards of the breed explicitly state that excessive coat is to be penalized! It also states that a cottony coat shall be penalized. (Wonder what the American Cocker Spaniel Club considers cottony?) On the head the hair is short. That's what the standards call for. It does not say; clip the hair on the face with a number 7 fine tooth clipper blade. It does not say to purchase a pair of thinning scissors and shape the entire dog to fit the standards. The official standards do say:

"Medium length, with enough undercoating to give protection…but not so excessively as to hide the Cocker Spaniel's true lines and movement or affect his appearance and function as a sporting dog. The texture is most important. The coast is silky, flat or slightly wavy, and of a texture which permits easy care. Excessive or curly or cottony textured coat is to be penalized." (Glover, 70)

So what do you think? If you were a Judge, would you ever pick this breed to represent the best of all breeds when it doesn't even follow its own standards? And how about going hunting?

Take a look at that noble long lost American cocker! If I had that dog, I would be in dog heaven! But there is no such dog any longer! Here's the new American Cocker Spaniel. What do you think? Should we believe the American Kennel Club when it states that its main goal is "to do everything to advance the study, breeding, exhibiting, running, and maintenance of the purity of thorough-bred dogs." (American Kennel Club, p.1) By the way, that was a quote from one of their pamphlets. Do you see anything wrong with that statement? They called "purebred dog's" "thorough-breds". That's a horse.

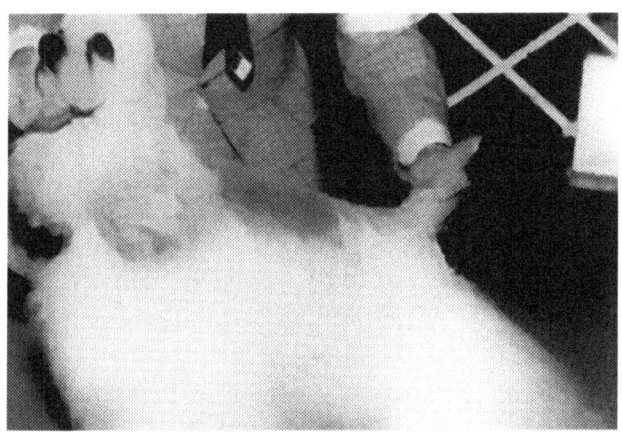

Cotton Coated Show dog.

Going forward, there is nothing wrong with creating this beautiful new breed of cocker spaniel, it is a great dog, if that's what you want.

Today because no one declared a new breed for this new cotton-coated Cocker Spaniel and, therefore, did not protect the old true American Cocker Spaniel, that old breed of spaniel is all but lost to the world!

The formation or creation of breed clubs and breed standards are the true beginnings of pure-breeds. If the breed standards are well written and the true lovers of the breed stand up to any persons or conglomerates in the protection of their dogs then a true breed might just stand a good chance to keep its nobility. Thus, the lovers of the breed would keep their dogs.

Before written standards or guidelines appeared, breeding one dog to another was sporadic. Successful breeding was done only on the dog's abilities to do his masters chosen line of work. The farmer's and Indian's attitude on the subject of dogs was that the dog was an extension of his family and was a useful working animal that either did a job and retired or was used for food for those that could or would provide for the family unit.

Here are some histories of pure breeds that you might know:

Doberman Pincher —- In 1870 a man named Louis Doberman of Germany breed the Rottwieler with the German pinscher. He also bred into the line a black and tan English Terrier and one of his shepherd dogs. In return he produced a dog they called the German Terrier, which soon became the Doberman Pinscher. This is why if one breeds a Doberman to a German Shepherd the pups come out looking like short hair German Shepherds with half cocked ears.

The Dalmatian —- This dog was originally a guard or war dog in Dalmatian or Croatia. There is a possibility that the Dalmatian was being confused with the spotted German Mastiff, which was known at one time as the Tiger Dog. The Dalmatian was probably a smaller version of the harlequin Great Dane, who was bred to the pointer to reduce the size and improve the markings.

The Boston Terrier —- This dog is a cross between a bulldog and an English Terrier. Careful introduction of French Bulldog blood helped in perfecting the breed.

The Bichon Frise —- This dog descended from the Water Spaniel or Barbet and existed at first in four forms. (The Maltais, the Bolognese, the Havanais and the Tenerife.) In 1933, an official standard was adopted and the breed became established under two names. The Bichon and the Tenerife. It was Madame de Leemans who proposed the name Bichon Frise.

Yorkshire Terrier —- The creators of this breed probably took the Clydesdale Terrier as the starting point then introduced the old English black and tan Terrier. The Skye and Maltese would give the length and silkiness of the coat. The Dandie Dinmont Terrier may have also been used.

Silky Terrier —- The two main ancestors of this breed were the Yorkshire Terrier and the Australian Terrier. Breeding the Skye, the black and tan and the Maltese produced the Yorkshire Terrier. The Australian Terrier was produced from the Skye, Dandie Dinmont and the Australian Terrier. In 1900, in Australia, the Yorkshire Terrier and the Australian Silky shared the same breed club. Soon after the turn of the century, breeders in Sydney started the Sydney Silky Club and the breed became widely known as the Sydney Silky. It was not until 1959 that the Australian National Kennel Council adopted a standard for the breed and changed its name to the Australian Silky Terrier.

Shih Tzu —- The Lion dog of China, Tibetan Spaniel, Lhasa Apso and the Pekingese made up this dog. In the 1930's, these dogs were classified all together in one bunch as the Apso's. They arrived in the U.S.A. registered as Apso's. In 1940, the breed was granted a separate registry in Britain.

West Highland White —- All the terriers of Scotland came from the same rootstock. They were all once the same dogs. When a litter of pups produced a white or a tan pup, breeders would breed the whites and tans to others with the same colors. Sometimes a longer legged pup would pop out in a litter and a man by the name of Col. E. D. Malcolm of Poltalloch would go on to bred those long legged terriers together to come up with the West Highland Whites.

These are just a few. As I stated before, all dogs were selectively bred at one time or another and a new breed was formed.

Q- How many generations make a purebred?

A- As many as necessary until all prodigy or offspring beget offspring that resembles each other.

What is a wolf dog? —- Though scientists today categorize the wolf and the dog as the same species, I cannot. To me a wolf is not a domesticated animal nor can it be, in my opinion. The wolf is a wild animal with extremely strong inherited instincts. I do believe it is dangerous to think of wolves as just untrained dogs. If wolves were left wild they would continue to be a pure strain.

The biggest reason for me to not categorize the wolf and the dog as the same species regardless of their microcondria DNA is that the female wolf and the female dog do not even share the same estrogen cycle. I don't know about you but that's an awful big difference between the two in my book!

I have done a lot of reading on the subject of wolf/dog crosses and as of today I personally do not believe that a wolf will breed with a dog without the tampering (artificial inseminations) of the pair. I prefer to believe the old writers who have continuously written in their experience that no wolf would breed with a dog, as they prefer to eat them. I also would believe a native Alaskan Eskimo before I would anyone who writes fictional stories of wolves breeding with dogs. As for the facts that wolf/ dog crosses do exist, I believe that they do. But what kind of pet would that be? I do not think that I would want to take any wild animal out of its natural habitat just to please myself and my ego.

What is a Pedigree — A pedigree is a lineage, a written log of ancestors, or a family tree. Most people think a pedigree is a written log of a specific breed with mothers and fathers within that same breed and only that breed. That thinking is incorrect. Again, a pedigree is a record of offspring and mating's or marriages. If a person keeps a written record of all breeding's (even of inter-breeding's between other different breeds) you would still have a written record of mating's. That is a pedigree.

Pedigree - 1. A line of ancestors: lineage 2. A list or table of descent and relationship, esp. of an animal or pure breed. (Funk and Wagnall's, p. 483)

Genealogy - 1. A record or table showing the decent of an individual or family from a certain ancestor. 2. Decent in a direct line from a progenitor; pedigree. 3. The study of pedigrees. (Funk and Wagnall's, p. 266)

Some folks ask, "How can the Shepalutes have a pedigree?" The answer is simple. I have recorded all the mating's.

I used mostly registered dogs for breeding's that have their pedigrees go as far back as the beginning of the registry club they belong to. The one difference is that I know the breedings of the Alsatian Shepalutes, because I bred, recorded and have taken photographs of all my dogs. I do not know for a fact that the purebred animals I have purchased (with their grandeur lines of champions) were who their breeders claimed them to be.

A.K.C does not require photos and/or tattoos. I have heard that they are introducing DNA recordings. Smart.

Inbreeding or Line breeding

It really upsets me that so many ill informed people including doctors of veterinarian medicine use this word (inbreeding) to describe something terrible, disgusting and almost alien.

Every time someone refers to the depletion of the breeds, whether it is the coat or skin allergies, they refer to this word "inbreeding" as if this were the problem to end all problems. They may be correct in the way breeders chose to improperly inbreed dogs, but when used correctly inbreeding sets genetic characteristics.

Inbreeding is a good thing to do if the breeder understands what he/she is doing. The real trouble here is that a lot of breeders do not know how to inbreed.

Let me explain how to inbreed, or line breed. You have to have guts, determination and perseverance and you must go the distance! Most of the problems in the breeds today are that breeders didn't/couldn't go the distance. Kind of like riding a jumper and when you get to that 6-foot wall you pull up and slam into the wall.

Inbreeding is breeding father to daughter, half brother to half sister, son to mother, and by the closest inbreeding of all, brother to sister.

Inbreeding concentrates both the good features and the faults. It strengthens dominants and brings the recessives out into the open where they can be seen and evaluated. It supplies the breeder with the only control he can have over the combining and balancing of similar genetic factors. Inbreeding does not produce degeneration it merely concentrates weaknesses already present so that they can be eliminated.

Do not inbreed unless you have a really great dog. That is the difficult part. Getting yourself a really great dog and knowing it inside and out. But, let's say you have a really great stud dog. You have stepped back and have looked at this dog objectionably. (In my opinion here lies the problem. Everyone thinks his or her dog is the really great dog!) I don't blame you; we all want to think that, but can you look at your dog objectionably? As you can see this may be a slight problem for a lot of folks.

Let's say you know your stuff and your dog is exceptional. You check out his pedigree. Let's say his pedigree or family tree is not a lie. Next you contact all people who have all these dogs listed in the pedigree as far back as you can go. You go to see the dogs, take pictures and spend some time with them. You look at the family photo albums and you get to know the dogs. You take notes and color-code their faults on a chart that you have made up for all the dogs that you are going to see. Next you go to all the aunts and uncles in the direct upline and color-code their faults. Then you go to all the brothers and sisters you can find and color-code their faults. What do you mean you don't see any faults? Hehe. Is there a perfect dog? Call me. Just joking.

Ok, now you have got an Idea of the genetic inheritance of the coat, skin, eyes, ears, body structure, bone density, x-ray and hips. The character of the whole mass of who your stud dog is and of what he possesses genetically. You will be able to see into the puppies that come from him which characteristics they have from these lines and what characteristics each of the parents are strong in.

Do not rely on championships handed-out in a show ring by judges who may be prejudice or who may have gotten a kickback when putting a certain dog up. All littermates could be dysplastic, rely on your investigative research. Show breeders may think that having a champion dog means that the dog has fewer faults than the other dogs and will breed fewer faults. It sounds logical, doesn't it? Their only proof that they bred a great pup would be if the pup wins in the show ring. What If I had twenty-six dogs and put them all in the show ring? Who would win? What if I found out that another dog might win over mine, so I kept all of mine out of the ring at that time? Can you understand why this type of thinking may deteriorate the whole breed? A champion stud dog, with hidden recessive bad genes (for say "hyper ness"), is now breeding hundreds of bitches throughout the world. Do you think that a judge might possibly see that hyper ness in that dog and not put that dog up? What if hyper ness was confused with loving the show game and presenting himself as a happy go lucky dog?

Canines have been having pups and breeding on there own for longer than humans have been on this earth. The elements, the pain, the suffering and the dying are God's breeding plan and his breeding has made the wild canine a strong group of mammals. Perhaps that is one reason the mutts of the world live longer than our pure-breeds. (And cost less in veterinarian bills.)

Here is my suggestion: Plan your kennels with grass, dirt, rocks and a stream going through it then leave the dogs to breed. Feed them good natural foods. Plant some grapes along the kennel run. Put some mint next to the water trough.

Test the pups and write it down on their charts. The first 6 weeks you should be observing and reporting all that these pups do. Who whines? Who walks first? Who eats more than the others do? Who is too weak to grab on to the teat? Of course some will die, especially if there is a lot of weakness in your breed of dog in the beginning.

For Alsatian Shepalute pups:

First test: For the first 3 1/2 weeks do not touch the pups! This is most important. Do not let them smell you or know you. Figure out per your breed standards what your pups should be like and with the least bit of interruptions that might sway the true nature and character of the pups, do your first real test. Remove the mother and lock her out away from the pups. Go into the pups and put your hand on the floor about one foot away from the pups. Don't say a word. Who comes to investigate first? Who runs? Who cries? Who shakes? If you have any like the before mentioned characteristics and that pup does not fit the breed standards what are you going to do?

Breeders belong to a strong unity of humans who understand the great importance of the work they do. They realize that every dog that they bring into the world is a representative of their dogs and of that particular breed. Breeders are a unique type of person and not everyone can be a breeder. It's not easy and we do cry a lot at times.

Inbreeding is the way to fix certain traits. I had to inbreed three times strong back to a particular stud dog. Then I bred to the bitch's half brother. Breeding like this brings out any hidden recessive traits that no one sees. It doubles and triples up those genes. That's how I got rid of some really bad problems that came from the registered purebred stock.

Inbreeding also makes all of the dogs in a family so similar that they would appear to be twins. It sets the genes. This is why I continue to tell you that when inbreeding, the breeder absolutely must select only the strongest and most prolific pups in every litter. And it may not be the prettiest! So how do you know which one to breed back too? You must leave the pups do for themselves. You must observe and report and not touch the pups. You must do repetitive tests on all pups at certain ages in their growth. You must be strong in your goal and you must not waver off the path.

Inbreeding and line breeding are essential in establishing desirable traits and weeding out the undesirable. So when a person attacks breeders with that very bad word "inbreeding" and says that inbreeding is deteriorating the breeds and causing weakness and allergies in the pure-bred strains, then I say you do not know what you are talking about when it comes to the art and skill of breeding. I will agree that uneducated breeders may be inbreeding or line breeding and not going back far enough to double the bad traits and weed them out. I firmly believe that educating folks is the way to go, not bombarding breeders with the news media and unleashing the gates of hell on them. That just spreads hate and discontent and we have enough of that in the world.

A breeder must also understand the hybrid vigor of out crossing and use that when needed. The mirror opposite of inbreeding is out breeding.

Education is the key to understanding and understanding is the key to the universe as love and understanding hold us all together.

Line breeding —- Line breeding is a broader kind of inbreeding that conserves valuable characteristics by concentration and in a general sense gives us some control of type but a lesser control over specific characteristics. It creates "strains" or "families" within the breed itself, which are easily recognized by their similar conformation.

Line breeding entails the selection of breeding partners who have one or more common ancestors in their pedigrees. These individuals occur repeatedly within the first four or five generations so that it can be assumed their genetic influence molds the type of the succeeding generations.

Selection is an important factor here also, for if we line breed to set that specific type of character, then we must select breeding stock in the next generations, which is the prototype of that individual.

1. Decide what traits are essential and what faults are intolerable. Character and temperament must be included. You must be able to strip away the training influences to see the raw dog.

2. Develop a scoring system and score selected good and bad faults with your breeding aim in mind.

3. Line breed consistently to the best individuals produced which by your temperament and character tests show that they will further improve the strain. Understand that coat and coloring of any animal is the easy stuff to fix. The hard part is the unseen character of the dogs.

Outcross breeding —- Out crossings can be made to bring in the wanted characteristics if they are missing from the basic stock. It is also vital in strengthening lines. Most breeders refuse to outbreed because the result might be no ribbons, recognition or monetary rewards.

Relationship need not be close in the foundation animals since wide outcrosses will give greater variation and therefore, will offer a much wider selection of the desirable trait combinations.

Outcross breeding is the choosing of breeding partners whose pedigrees in the first five or six generations are free from any common ancestry. One of the partners should be inbred or closely line-bred. Out crossing will bring in new and needed characteristics into the strain. This is vitally important for vigor, strength, and vitality in your strain.

Breeding for Intelligence

What is intelligence and how do we define it when we talk about breeding dogs?

Funk and Wagnall define intelligence as "the ability to adapt to new situations. Understanding; reasoning." If this is the true definition of intelligence, then animals cannot be classified as intelligent. I beg you to reread that definition of intelligence. The Homo sapiens are the only creatures on earth with the ability to reason. That is one of the ways that we were formed in the likeness of God himself. I have not found any animal that could reason in my lifetime. So what is intelligence in the dog?

If by intelligence in the canine we accept the definition "the ability to cope with the environment" then God certainly has the upper hand here. If a canine cannot cope in the wild, it dies. What about domesticated dogs? Certainly most of the breeds would never make it in the wild. I guess all our breeding's for the perfect domesticated dog took the intelligence right out of the canine!

I believe that any doggy intelligence tests given to different breeds of dogs could only be validated if judged within the separate breeds and amongst themselves. Certainly, a Chihuahua cannot compete with a German Shepherd Dog when a human wishes to determine which breeds of dogs are intelligent or not. It is rather stupid to judge "the ability to cope with the environment" of a dog in a test of all breeds tested together. But it was done! And it was on National TV. I knew right away that the person performing such a test could not know much about the domesticated dog breeds throughout the world. Different breeds have been bred for different characteristics. And tell me who would want a domesticated dog that was closest to the wild? "Able to cope with the environment." That's the reason civilized man took the dog out of the wild, to domesticate him. To make him stupid? Ha! No way! The reason was to make the dog fit into our domestic ways, to take the wild out of the dog. And we did it with the most submissive of animals for they are the ones that were able to fit into the Homo sapiens world, to take care of his clan, to hunt and to stay by his side.

If Intelligence in the canine is defined as "the ability to adapt to new situations within the earths wild environment and survive" then by this definition alone the wolf would come out as intelligent and I would not be able to use the word intelligent together in the same sentence with a domesticated dog. But, if someone included the definition of the domesticated canine as: "the ability of the species to perform the work it was created for and to adapt itself to the changes in the environment and the conditions it encounters as it performs its designated tasks", then I believe we could once again use the word intelligence within the breeds.

The easiest way I know of to provide such breeding for intelligence tests would be within the breeds own club. Yes the breed club! Who would know more about that particular breed than those who admire and love the breed? Those who have given their lives to the breed and who have studied bred and tested their dogs.

How a dog thinks

Animals do not think as you or I do. They learn by a process of what works and what doesn't, combined with genetic instincts. God created animals with a sole, if you believe in the words of the bible in Genesis. Unlike humans though, they do not have the ability to reason. Their thinking process is just a learned process by the way of what works or what doesn't work for them to survive or to get what they want. Many folks will disagree with me and bring up many examples, far too many to write in this book. If you take an open and objectionable look at any of those examples you will find that the animal learned the behavior that you are describing.

Animals do not plan ahead, nor do they know what "time" is. When they go to the door at five o'clock to meet you they have learned that the consistent sounds of the day tell them what is going to happen next. School busses, mail trucks and the morning alarm all signal them. It is because of consistency. That is learned, they know what will happen next because it always does.

They act in the "now" and they have a learned process of what they have done before that is applied to what they are doing. They run off hunger and sorry, but they do not know what love is or means. After 18 years of being with you they do miss your presence around them. They have grown accustomed to you always doing what you do and being an entity within their life. They do know that they get a pet if they are in your lap because it's happened before. If they look at you with a head cocked they have learned that it brings a desired result. They read your body language. They do learn words by the sound or the way the word is spoken to them through learned experiences. If I say to my dog "you bad dog, you are so ugly and you stink and are so stupid" and if I say that in a cute, high pitched voice with a smile and I bend so that I am less a threat, the dog will wag his tail and come to me happy and excited. They do not know words unless you teach them. It is the tone of your voice and your body posture or what I call the universal language.

Of course they know pain but they are not in their pain. If something hurts it hurts them for as long as the pain is there. The greatest thing about animals is the way they adapt or except their surroundings. They make the best of their lives by becoming as close a part of it as they can. They do not know or think how they can change their lives. They are in the "now" always.

One of the biggest reasons that I know of for you not to be able to understand or communicate unconsciously to your animal is your state of mind. Many, far too many people think of their dog as having humanistic reasoning. If you must think of your pup as human, think of him as a 6-month-old child with autism. Do not get upset if your pet does not live up to your human way of thinking. Your dog is not a human and never will be. Dogs do not harbor gratitude. They are not thankful. Your dog does not have a sense of duty to you or to any other creature than to himself.

Breeding for Charm

The vast majority of American dogs are kept as pets. That is not an understatement. Guns are all but extinct as is hunting and soon to be fishing. Along with them go the true bird dogs, the hunting dogs of yester year. The only things keeping them alive are the field trials. Yes, the Homo sapiens of the new millennium are city folk and companion dogs are in high demand.

Breeding for charm is not true breeding unless one construct's a standard for this charming little dog and then breeds dogs to create more charming pups. The majority of the public does not care about pedigrees. The public wants cute charming puppies. A cute puppy goes quickly. But what is considered charming? Some folks just breed cute little mongrel charming pups. The problem here is that the characters of the pups are not a sure thing. It is not predictable, not stable and neither is the charm look. And of course, what is charming to one may not be charming to another. And within this, I must also say that charming pups are numerous. But what about when they grow up? Perhaps that charming pup won't be so charming?

The problem here is that in order to become breeds, dogs must breed true. That means that all pups coming from these breedings must produce like pups. Then in each generation from that, all pups produced would also be charming per the standards of what one would consider charming.

Genetic Traits of Canines

There are a lot of variations of the totality of the complete dog that must be acknowledged before you can say that one thing is more dominant than another. There are so very many different factors that must be considered. Even the same pair of genes in identical twins produces a slight variation. Therefore, a breeder must keep in mind the variations of all the genes.

Keeping that in mind, I will round off what I have found to be true in my breedings.

Coat Length and Density

Once I bred a short-coated Alsatian Shepalute to a short-coated purebred Mastiff and it produced F-1 pups of a thick, medium-length coat. Twelve puppies were born in the litter, none died and none were short-coated. The muzzle of all the pups was longer and the bone structure was thinner than what the Mastiff parent had. The ear placement was not attractive and all ears were semi-down. The dominant up-ear had affected the hang-ear according to my varying slide rule. The muzzles were black along with the tip of the tail and the rims of the ears. The eyes were brown like the Mastiff, not yellow. They resembled Saint Bernard's. The F1 pup was bred back to an Alsatian Shepalute, which then produced short coats.

Other breeding's — A dense short to medium-coated Malamute was bred to a short thin-coated German Shepherd Dog with a double coat.

All the pups came out with a thick, rich coat of medium, short length. All pups had light brown eyebrows. The saddles were larger (extending down the sides of the legs) and blacker.

I crossed a German Shepherd (black/tan) with a Siberian (red/cream) and got all wolf looking dogs which were extremely unmanageable. All were beautifully marked with nice short thick coats! The litter had two red/brown, three black/tan, 4 tri-colored (all dogs carried the graying gene which is what I wanted) the dark pups turned lighter with age.

The eyes were light brown to yellow. There were no blue eyes. But, they were very hyper and did not pay attention, nor did they care for training. They did not learn quickly, but rather took a long time to train. Over and over again I told them that whining was an unacceptable behavior. They did not get it, nor did they ever.

Note: The Large body is recessive to a thin body but short legs are the dominant with the large bodied dogs. Short muzzles are also dominant with a massive body. They go hand-in-hand. Therefore if you breed for girth you will loose height. If you breed for height you will loose girth.

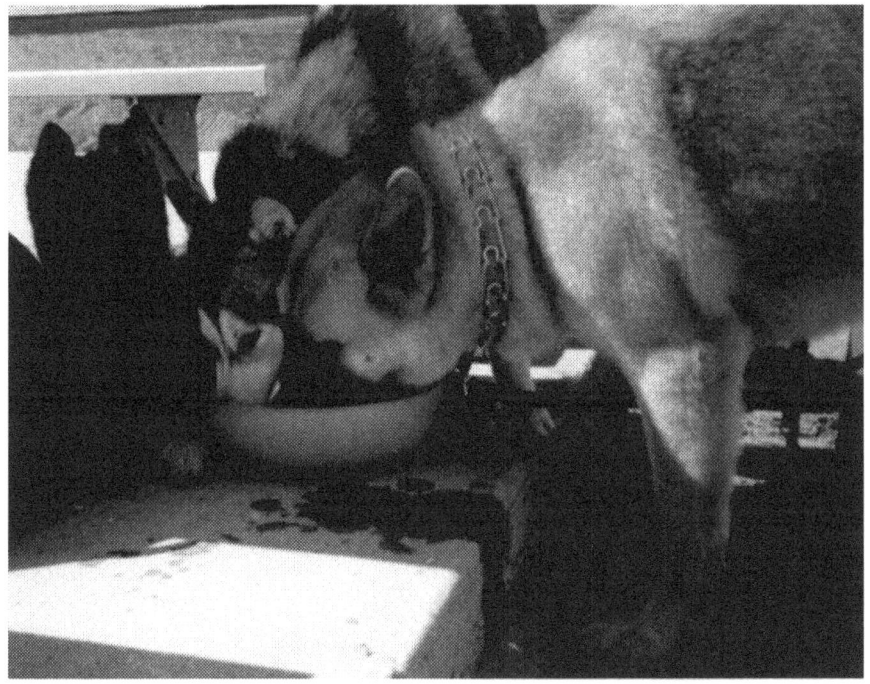

Left to right: MalxAkitaxshepxmastiff, Mal, Akitazshepzmal

Dominant	Slider	Recessive
Ears		
Up ears	weak/half up	Drop ears
Legs		
Long	Medium	Short
Bone Density		
Thin	Medium	Thick/Large
Hyper ness/nervous	mild	Lacksidasical
Muzzle		
Long/thin	Medium	Short/Thick
Tail		
Curled	Variations	Drop/hanging
White Patches	Variations	None
White tipped Tail	Variations	None
Black Tip Tail	White	Body color
Long coat	Variations	Short coat
Feathering	Variations	None
Bark		
High-pitched	Talking	Non Barking
Digging		Non Digging
Jumping		Non Jumping
Dog Aggression	Variations	Gentleness
Howling		Non Howling
Sensitivity		Hardness
Prey-Oriented		

Tests were preformed on a litter of Shepalute F-3 generations in comparison with the Siberian Husky x German Shepherd Dog litter of the same age. The Siberian mix consistently whined and howled. That behavior sparked up a barking in the Shepalute litter, of which I consistently slapped a newspaper on the wooden beam that held these pups in a marked off area. The Shepalutes only needed two corrections, showing them that the behavior was not acceptable. They then curled up and went to sleep.

Another test was also conducted at the same time. A wooden bar four inches tall was the only barrier holding the pups within the appropriate designated space. The Siberians consistently climbed out. The Shepalutes never did. When a Shepalute put his paw upon the beam a newspaper was slapped against the beam and the pups paw replaced inside the pen. The same was tried with the Siberian mix. It did not faze them.

Owners were found for the Siberian mix pups. They were informed of the character and the breedings. Each pup was sold with a copy of the papers of the breeding's and the pedigrees along with photos of the parents.

I crossed a small cream chow with the Shepalute. Beautiful pups with nice thick coats. Too bad they were all shy. They were not what I was looking for. All pups were sold to good homes. All new owners were informed of my breeding results and all offspring had charts.

Reaction to Vaccination or what I refer to as Touch Sensitivity —- All puppies that showed any signs of touch sensitivity were not used in the Alsatian Shepalute breeding program. After 10 years of breeding, I now have no pups showing these characteristics. When I introduced the English Mastiff into the lines, I again had to weed out the touch sensitive squealers.

Jumping and Climbing out of Kennels —-Shepalutes will not jump over a 4-foot chain link fence, but they will jump up into the bed of a pick-up truck when giving the command and are excited to do so. I know they can jump, but they will not attempt to jump over the 4-foot fence not even to get at the chickens or goats. I have some Shepalutes that will go under a fence but the drive to do so is not high. I have had Shepalutes who were hyper and jumpers but I did not breed to them. Once in a while a pup will still be born who will be a hyper dog and that dog will jump on top of the vary kennel that is in the main kennel run. That particular dog's physical look out weighed the hyper ness only because of the ratio factor. Only the mellowest of pups from her litter were chosen to continue in our breeding program. Breeding to that mellow pup out of that hyper female will not give me complete mellow pups in the next generation. I will need to continue a strict and intense search for only the mellowest of those next generation pups. Mellow pups are very hard to come by because the hyper ness of the canine is so very dominant.

Yellow eyes —"The Alsatian Shepalute shall have a secretive yellow-eyed stare."

The yellow eye was among the last characteristics that I concentrated on. I had a Shepalute that had blue eyes until the age of 5 months. This worried me greatly! Where did I get this blue-eyed pup? I had been breeding the Alsatian Shepalute for over 10 years when I began concentrating on the yellow eyes. I had bred many German Shepherd Dogs into the lines. Only two German Shepherd Dogs that I used had a light eye, although, the Mastiff I used had a slightly light colored eye. (It could have been darker for show.) All the other dogs had deep brown eyes. Where on earth did this blue-eyed pup come from? This pup was the third generation out of the Mastiff and the fourth generation out of a yellow-eyed Shepherd. The only thing that I could think of was that this particular breeding caused a pull from a gene that was in one of the beginning dogs long before any of my dogs existed. Not so! The next few months the blue eyes gave way to a beautiful yellow eye!

I went on to inbreed this male stud dog to produce the great yellow eyes of the Alsatian Shepalutes. That was easy compared to the shyness or touch sensitivity and hyper ness that took so many years to get rid of. I am still finding pups too hyper for my liking that are popping up in some of my litters.

When a truly dominant trait is eliminated it will never show back up, it is lost forever.

Chapter 3

Creating New Breeds of Dogs

Breed Categorization

Why bring a new breed into the world?

In the dog breeds of today how many dog owners' use their dogs for what they were bred for? Today's dogs still possess those inherited instincts. In my many years of dog training it has always bothered me that owners were conditioning breeds of dogs that were bred for a specific purpose, into becoming a family pet. I have seen them try to modify the pup's behavior. These dogs were placed in city dwellings and were suppose to fit?

There are dogs that were bred specifically for that purpose. They are classified in the breed books under the category "toys." The toys are the only breeds that were specifically bred as a pet.

Instead of changing old breeds into a new breed of dog that is in demand today by breeding natural instincts out, I felt I should create a new breed of dog. A plain ol' companion dog.

The American German Shepherd is not the same dog as the German Shepherd Dog. The American Shepherd is more of a companion dog. Since the American Shepherd is still registered as a "German" Shepherd Dog it is suppose to be a "working dog" as described in the country of origins standards of the breed. Working dogs were bred to do jobs such as guarding life and property. Well, this tends to confuse the public as to the proper character of a German Shepherd Dog. How come a person could have a great Shepherd for twelve years and then when they go to purchase another Shepherd they find it intolerable?

Hunting breeds are being bred as companion dogs also, as in this day and age, that is what the public wants. Owners who want to enter their dogs in field trials find the breeds can not perform as they use to because breeders are selecting the qualities that the average person wants. This is causing a separation within many breeds' characters.

By creating a new breed of dog and classifying it as a companion dog, the public would be able to use the categories that go along with the different breeds of dogs and the old hunting dogs would be protected from breeders who breed family pet dogs.

As things are now the public doesn't understand what those categories are. When choosing a lifetime pet it would be so much easier to use those categories described in most breed books to find the pet that best suits your family. Have you ever tried to find a family pet by looking in the different breeds classifications as to which breed would best fit your lifestyle?

Let's take a look at the breeds and their categories:

Group #1 Sporting dogs: gun dogs, bird dogs. These dogs hunt; locate and retrieve game birds.

Group #2 Hounds: for hunting large game, hounds for hunting small game. Coursing dogs. Dogs that see their game and chase it. Trailing dogs. They hunt all game but birds.

Group #3 Working dogs: these dogs work for people doing many different jobs. Guarding and draft (pulling) dogs.

Group #4 Terriers: the working terriers, terriers that hunt and/or go to ground. They dig, bark and run after rodents.

Group #5 Toys: companion dogs, bred mainly for pets. Small pets.

Group #6 Non sporting dogs: and all other dogs.

Group #7 Herding dogs: these dogs herd or gather up farm animals. Drive or move animals in a group into a designated pen or area.

Some of you probably never knew this categorization existed. Well, know you do. Kind of makes it nice to be able to see which breeds were bred for what and makes it a bit easier for us to choose that perfect family pet.

Please keep in mind that for hundreds of years these breeds have been breeding within their own breeds and categories so their instincts are set very strong. A hunting dog was bred to hunt. Terriers were bred to go to ground (dig, bark, bite or kill their prey). Herding dogs were bred to herd animals and so on. To expect a true working dog to fit into a family companion dog group would take years of re-breeding to weed out the working dog that he is.

Around 1570 A.D., the Caius College in Cambridge, England put out the first listing or "groupings" of dogs that I know of. All grouping of dog breeds come from there. I am using the groups that the major dog registries have settled upon. I did not make up this list.

In studying these groups I see two groups that fit into the category of a companion dog. (Toys and Non sporting dogs).

If you live in the city you will find that there are many rules and regulations concerning your dog that you must live your life by. Each city, town, or state's rules and regulations are different. You may find these rules or laws in your library. Some of the major regulations are in this book's chapter on "Dog Laws."

If your dog is a nuisance you may get sued or your neighbors might start doing little things to upset you. I think they do that because they are extremely upset at you and your pet and do not wish to confront you. Maybe they have yelled at your dog or called you on the phone complaining about your dog. Maybe you just start finding cat feces on you're front door step or your car gets keyed. Now there becomes a feud between the neighbors. Someone may go to jail or end up killing something as has happened recently in the news. I felt very bad for the neighbor who was being so harassed that he struck out his foot at his neighbor's little dog and ended up killing it. Of course it wasn't the dogs fault. And I am sure the neighbor did not have "intent" to kill the little dog who suffered the worst because of what his owner had made him do, putting him in harms way like that.

You can see though that it would help if a person's dog had certain qualities. Qualities that would not tend to enrage the people living near you. Note that we are not talking about the owners here. I do realize that everyone needs to use their common sense, but some folks just don't have any. (I guess that's for another book). What we are focusing on in this chapter is the best breed of dog that would fit into our society or neighborhood (as it is today) with the least amount of modification or training.

Ok, so let's take a look at the two groups we have to pick from above. I am sure there must be more dogs in those other groups that would make great companion dogs also, but a hunting dog is and should be a hunting dog. To require a hunting dog to be a companion dog would mean that someone would have to modify that dog's behavior to calm him down enough to fit him into the neighborhood. And what if that hunting dog saw a squirrel? I'm trying not to change the behavior of a dog by breeding the hunting or herding instincts out of the dog and, therefore, loosing a great breed. Nor am I suggesting one modify the natural instincts of that dog to have him fit into a companion dog's role. Some would argue with me that a hunting dog could be a companion dog. No argument. It can be done. I'm talking about the many everyday folks who don't want to, or can't modify their dog's behavior. Folks who don't have the time to go to dog classes. Folks who want it simple. Folks that want a dog to just be itself and who don't want any hassles with the neighbors.

Let's examine Group #5, the toys and companion dogs:
Affenpinscher
Brussels Griffon
Cavalier King Charles Spaniel
Chihuahua
English Toy Spaniel
Italian Greyhound
Japanese Chin
Lowchen
Maltese
Manchester Terrier
Miniature Pinscher
Papillion
Pekingese

Pomeranian
Poodle (toy)
Pug
Shih Tzu
Silky Terrier
Yorkshire Terrier

From this list we have to take away all dogs who will disrupt their neighbors by their high pitched, uncontrollable and consistent barking. Barking dogs are the number one aggravations to neighbors who get so upset over the continuous barking that it drives them out of control!

Now, we must take into consideration that a good trainer can control a dog from barking while he or the family is around. One can also place a non-barking electrical collar on the dog while the owners are gone, but we are not talking training here. So, eliminate all the consistent barkers and you get:

Shih Tzu
Pekingese
Pug

These are all the rather calm companion dogs that can be more or less easily trained to be acceptable to the neighborhood. Any breed of dog can become a barker if the situation allows it.

Ok let's go on. Let's check out Group # 6 the non-sporting dogs:
Bichon Frise
Boston Terrier
Bulldog
Chow Chow
Dalmatian
French Bulldog
Keeshond
Lhasa Apso
Poodle
Schipperke
Tibetan Spaniel
Tibetan Terrier

And let's eliminate the barkers so we have a new list:
Bulldog
Chow Chow
Keeshond
Lhasa Apso

Let's just go ahead and include group # 3 the working dogs:
Akita
German Shepherd
Doberman
Rottwieler
Mastiff
Bull Mastiff
Boxer
Poodle
Saint Bernard
Newfoundlander

And let's eliminate the notorious barkers as best we can, so we have a new list:
Akita
Mastiff
Boxer
Newfoundlander

So here is your combined list of possible companion dogs that don't need to be modified in their trainings to genetically un-inherit their barking traits. The more quiet non-barkers of the list so to speak. Some of these dogs can become notorious barkers if allowed to do so but for the most part they don't bark much. It was very hard to choose but I had to make a choice based on what I have seen. So now here is the new list.

Bulldog
Chow Chow
Keeshond
Shih Tzu

Lhasa Apso
Pekingese
Pug
Akita
Mastiff
Boxer
Newfoundlander

Let's continue. What if you have frequent visitors or children? We must eliminate all dogs who would be a threat to children or adults who may visit your home, unless you put up fences and warning signs. Remember we are looking for the perfect companion dogs that will be the least likely to cause any problems in today's society.

Here is the new list:

Bulldog
Pug
Shih Tzu
Newfoundlander

Of all the many breeds in the world, for which there are over 4 hundred, one can see that there are a very limited number of breeds that would be a breed that would cause the least amount of problems in this sue happy country.

To be completely fair, I should at least add 8 more breeds to this list that are in the miscellaneous classes and/or who are recognized by other countries throughout the world. The total of dog breeds that would fit into the companion dog list that are easily trained, don't bark much and who are not a threat to others would then rise to about twelve. However, I will not include them here, as I do not expect this book to be a best seller and to be read by others that may be living in the many different countries around the world. Let's go ahead and check out these four breeds a bit closer?

Pug —- The little pug can be aggressive for its size especially to children who the pug regards as a threat as kids may step or fall on this little guy. If the pug lives with an older couple the pug will not tolerate children, unless that older couple has grandchildren and the adults teach the kids not to bother the dog. You may have to put this dog in a vari kennel when the grandkids come over. When a pug is raised around children they do fine until they get old and cranky. Dogs that are small are not stupid, they fear larger things falling, stepping, or harming them. They will defend themselves.

Bulldog —- Now I know all dogs with teeth will bite and even old dogs without teeth can bite, but these three last breeds are good-natured and do not bark much. The one who barks the least in my experience is the Bulldog. They hardly bark and when they do, it's not loud. They also have small jaws, which is better than large jaws or better than a long jaw that can grab more skin. Short-jawed breeds do snore though. Hehe. The bulldogs throughout my lifetime have always had skin problems though, so watch for that as it can run you up a large bill at the vet's office. You might also have to put the dog down as the skin problems can get out of control.

Shih Tzu —- As a Groomer I have known an awful lot of Shih Tzu's! I have also bred them, as in my opinion they are the cutest of the little companion dogs! They also have a great temperament but are a groomer's nightmare when the owners of these little dogs do not bring them in to be groomed once a week! Yes, once a week! Unless you are a Groomer or hairdresser you will not be able to care for this dog unless you schedule regular appointments with your Groomer. It may not look like the dog needs to go in every week, but speaking as a groomer and breeder, unless you clip this dogs hair to about one half inch you will have tangles. Also if the breeder breeds the wrong type of coat this little dog may have tons of knots throughout the coat so you must comb this dog's hair every day.

I need to say that I have also seen mean little Shih Tzu's, but luckily they have not been many. If this breed gets popular as a small indoor companion dog and homeowners breed them without the help of a breeder, then we will begin to see many faults come out in this breed.

As far as my opinion goes, this little indoor companion dog is just the ticket if you want to groom and take care of a small dog. Know your breeder if you decide to purchase one of these. Ask about the coat quality.

Feel the coat and run a comb through it. Of course you should see the mother and father of the pup and you should be able to pick the dogs up with their tails a waggin! In other words do not get a mean line of this breed.

Newfoundlander —- This is a large, heavy-coated Northern dog. Similar to the Saint Bernard, but with a gentler nature and they are good with children. This dog was bred as a draft animal and can pull a child's sled or wagon. He has a great personality as long as the breeders continue to breed the good-natured character. He does take a considerable amount of grooming though.

Back to the Alsatian Shepalute —- Now the reason I started this new breed of dog is that I wanted a great companion dog. One that I could depend on in a crisis where I felt I had some type of protection with the least amount of trouble from the neighbors. A dog the neighbor wouldn't kick.

I honestly felt it was time to breed a dog that would fit into today's world. Not to change a breed to fit, or fight with a breed to fit. It should just fit. New owners who have never had a dog before that have purchased Alsatian pups were absolutely amazed at how quick these dogs learn. They are the most intelligent dogs that I have come across in my lifetime. So easy to communicate to. The most impressive character that even strangers comment on is the calmness of this breed! As I have stated in the breed standards, that calmness is the most important aspect of this breed. It shall not be hyper or shy! It is a constant chore and challenge to always breed better than the last because of those two dominant genes.

Now, there are a few different ways to create a new breed of dog, let's examine them.

Breed Crossing to Create New Breeds

This is the easiest way to create a new breed of dog. In this example, one would simply take a breed of dog and breed it to another breed of dog. There have been a lot of breed crossings to create new breeds. Take the Shih-tza-poo, the Cock-a-poo, the Terry-poo, and all the other "poo" crosses. They have their own registry and clubs and one day they may join the ranks of true breeds! I myself have even bred those little breeds and the public has loved them!

Breed crossing has been done over and over again, just for the fun of it, or just for the outcome. The starting of those new breeds didn't begin with an idea for a standard of the breed. The starting of these mixes started with the breed crossing of purebred registered dogs.

The reason one would want to know the dogs in the lineage is so that the buyer can get an idea on how the pup is going to behave or act when the puppy matures.

The difference with the Alsatian Shepalute breed is that the idea came before the choosing of the breeds that I felt would get me to that standard or blue print. This breed started with a breed standard and then dogs were bred to dogs that would produce what I was looking for genetically. As I bred generation after generation, I knew I would need to introduce a third breed into my lines. One that would be necessary to get those required results. These other folks who have created new breeds accepted the results of their breed crossings and wrote the standards that fit the outcome. Maybe they didn't even write a standard but just started registering or numbering the dogs.

New Standards for an old breed

We've talked about this one, how the Americans have always taken dogs from other countries and brought them here, then made up their own standards calling the new breed, the American Cocker Spaniel for-instance. The American Cocker use to be the Cocker Spaniels of England, which were much larger dogs. When the cockers were introduced into America they were bred smaller for the smaller game in this country. Then the cocker club decided on the name American Cocker Spaniel to distinguish it from its cousin, the English Cocker Spaniel. That's how it's done.

Now, I have seen a distinct difference between the American (German) Shepherd and the German Shepherd Dogs one sees in Germany. I have found the American Shepherds to be a far gentler dog than that of the Shepherds coming out of Germany. If one looks closely they will find that the American (German) Shepherd is taller with longer legs and not so hyper active.

Take a good look at the American-bred Rottwieler in comparison to the German-bred Rottwieler and you will see that the German Rottwieler has shorter legs and more chest. Even their standards of the breed show this to be true. The American standard calls for a 50/50 proportion of chest and height. In the German standards, they call for a 60/40 chest over leg or height. How can it be the same dog if the blue prints are different?

New Breeds by Mutations

Sometimes new breeds are created on the color of a dog alone. Take the West Highland White, for example. The Cairn, Scotty, and Westie all came from the same father stock.

Sometimes new breeds of dogs come from the mutation of coat consistency. The wire-haired, the curly-coated, the longhaired, the short-coated and the hairless are a few examples.

The Latest new breeds of dogs

I looked up "New Breeds of Dog" on the Internet and came up with 11,334 pages of listings on new breeds of dogs that you could go to.

I went to some of them and read down the list. I know most of the world's dog breeds and their names as I have studied them. I will admit though that it has been awhile. Well, out of all those many lists of dogs, here is a short list of breeds that I have never heard of:

Labradoodle
Kyi-Leo
Mc Nab
Mi-ki, toy
Mudi
Red and white setter
Russo-European Laika

Shiloh shepherds — I have read up on this one and I have studied all information pertaining to this breed as it peeked my interest. Some owners of this dog have e-mailed me pictures. Personally, my opinion is "how can this old lost breed just recently be found?" I remember seeing this dog when they first came out with the breed and the photos of some of the dogs resembled my dogs in a way. But looking again, they look like large, long-coated shepherds. That in itself is not a bad thing. A long-coated shepherd could be another choice for folks. Nothing wrong with that. I know some people that would like a long-coated shepherd. My problem with this old lost breed is that I refuse to have any dealing with anyone who lies and I just can't see how these folks can claim that this Shiloh Shepherd is a long lost shepherd that everybody overlooked. If they would have come out and just said, "We like the look of these big hairy shepherds so we are going to preserve them," that would have been fine. But, declaring them to be "the" lost shepherds of years gone by, I just can't figure that. Looking at the pictures of some Shiloh shepherds over the Internet, do you see any resemblance between them all? Do they have a breed standard? How come they all don't come out looking the same? I saw one of these dogs in person once and the owners swore by them. "Best dog in the world," they told me. I was very glad they loved their dog of course, but the dog didn't look anything like the Shiloh's I saw in the pictures. By the way, while studying the pedigrees of this long lost breed, I found my sister's purebred German Shepherd dog's pedigree. Boogie Down was a beautiful large boned shepherd with a great personality but he was hardly a long lost breed!

Toy Munchkin
Wolf Hybrid — This one I couldn't believe was on the list of new dog breeds.
Alaskan Husky—- We talked about this one.
Cane Corso
Cantel
Chart Polski
Chinese Foo Dog
Drever
Eurasier
Fell Terrier
Glen of Imaal

Miniature Boxers — I once read an article about a lady who was breeding Miniature Boxers. To this day I have never heard anything else about that new breed. I assume that it never made it. I did see pictures of these dogs and they looked as if they were a result of a one-breed cross between a pug and boxer with the pug being doubled in the next generation to keep the breed small. The dogs all had black faces and large eyes. The coat was the color of a boxer and the body looked somewhat like a Miniature Pincher.

Czech Terrier — Another article I read was about a new breed of dog that was bred by a scientist who stated he was looking for a dog to go into the foxholes. I don't really believe that a scientist did much hunting of foxes, but he could've. I mean when did he have time to fox hunt? But I will go along with the story. This new breed was called the Czech Terrier. Oh yeah, the scientist just so happened to be an all-breed judge and a Scottish Terrier breeder. (Hehe) The Scottish Terrier of old could have gotten into that foxhole, but in 1971 there weren't so many hunters as years gone by and as you may have seen evolution and dog shows seem to eliminate the old hunting instincts and body conformations. There are more dog shows and prize winnings for beauty and coat. Anyways, for whatever reason, in 1949 this guy chose to breed a Scotty with a Sealyham. Hmm? Both come from the same father stock. So it appears that the breeding of such dogs would be a "regressive breeding." A regressive breeding is what I refer to as a breeding of simplicity where the breeding results in a pull towards the natural state of all canines. That of the wild state. A breeding that goes backwards in time. Slender long-legged dogs with the dominant traits of the wild and old canines of millions of years ago. Again, it is a breeding that goes back towards the natural state.

Breeding's that have "evolved" are breeding's that resulted in specific breeds that hunt, curse, chase, tree, guard and herd. Breeding's that honed in on a specific trait or character.

This new breed of dog called the Czech Terrier was a true breed after only five generations and was accepted for registration by the International Federation of Cynology (F.C.I.) in 1963.

The Alaskan Klee Kai —- Linda S. Spurlin created this new breed of dog when she acquired a 17-pound gray and white female "husky" without papers. One of this dog's parents was "a small dog." The other was said to be an Alaskan husky. Now, what is an Alaskan husky? The Alaskan husky as I have researched is a group of sledge pulling dogs that do not have pedigrees or registration papers, yet. What I come up with for the Alaskan husky is that the name was tagged onto a group of dogs (slang) that ran the fastest when pulling the sledges. Color doesn't matter and neither does breeding as long as it produces a dog that can pull the sledge and win the race. By now perhaps, the Alaskan husky has a registration and breed standards.

Next Linda says she breeds her little dog, but to what? I don't know. I think perhaps in her web pages she hasn't made it clear. At least not to me.

Then, she gets a few of her brother-in-law's dogs that were the sire and dam of her first little dog. Her brother-in-law goes out of business and gives her the dogs. Linda says that the original Alaskan Huskies or the "Little Indian Dogs" are these breed's ancestors. She goes on to say that she added a touch of Siberian Husky and just the right amount of a "smaller dog." What that smaller dog was, she didn't say. She had thirty dogs and a friend of hers started taking pictures of the dogs and they began recording the breedings. She states a date that she sold her first miniature husky to a friend in 1987. They decided on the name from the Eskimo words meaning "little dog."

If you would like to learn more about this miniature husky, it's on the Internet. I am sure if you looked up the search word "New Breeds of Dog" you would probably come up with more than what I have here in this book. The reason I have included this chapter in this book was to inform my readers that creating new breeds is nothing new. It has been going on for centuries and will continue forever, that I am sure. The difference between those new breeds and my new breed is:

1. I do not declare this to be an old breed.

2. I have not registered the Alsatian Shepalute in any rare breed registry or other breed registry that wants to register any dogs it can get its hands on.

3. I planned this new breed before I bred any dogs together. It was not sporadic.

I do hope this chapter has helped some of you understand the dog world better. The big picture!

Breed Registration Clubs

I would like the readers to realize that breed registration clubs exist everywhere. They are formed by business people who created them enable to have a job, pay their bills, and perhaps make a little money above and beyond that of mere existence. (Ok, a lot more money).

Many times I hear folks ask me, "When will your breed get recognized?" I would really like the reader to give that question some thought, so perhaps I can help by asking some questions that will speed up this process.

1. Recognized by whom?
2. What is the purpose of being recognized?
3. Who would this breed go to for such salutation?
4. Who is important enough to be able to recognize this breed?

Recognition - 1. The act of recognizing or the state of being recognizing. 2. Acknowledgment of a fact of claim. 3. Friendly noticed salutation, attention. (Funk and Wagnall's, p. 555)

As the creator, of this breed I am not in favor of recognitions and salutations. I do not care for the mass quantities of awards and certificates that I may adorn my walls with. Who cares and what would be the purpose? To impress someone? And why would I need to impress someone or to verify what I do? To make something true? If something is, it is.

Let me tell you a little story. I started painting on china about 3 years ago and recently I wanted some of my work to be seen by the public, so for $80.00 I joined one of the many artist clubs. Now I can send the magazine a photo of one of my best pieces for their readers to see. For my $80.00, they sent me a beautiful certificate. So now I am a painter? In other words, for $80.00 you can be a painter too!

The only thing that matters to me is what is in front of me, who I see you as, not who the majority sees you as. It does not impress me that you or I can plaster our walls with cheep certificates! (Some are not really cheep I guess).

My opinions of all things are based on my experiences in life and not on the opinions of the mass majority. Nor do I categorize unless to be able to communicate my thoughts in a way others would be able to identify with.

I hope I have answered your question on "Are they recognized?" If not I shall make the answer more obvious.

1. They are recognized by themselves.
2. Their owners recognize them.
3. I, the breeder, recognize them.
4. They exist, therefore they are.

Every time a breed of dog is accepted into an "all breed registry club" that "all breed registry club" makes money and the potential for more money is ever present. That's the business.

Let's examine a fictitious all breed registration club. Let's say that each dog they register brings in 10.00. Each litter they register brings in 20.00 and each pup out of that litter (that the new owners register) brings them in another 10.00! Wait, that's not all! If they spend a little slice of all that income from the registration of over 300 different breeds, and they put on a big show in each state, they will make even more money.

At the show, each dog and each handler that enters in the ring must pay an entry fee and that brings in 8.00 each, just for the privilege of being there. At the door, you must get a catalog so you will know who, what, when, and where you are suppose to be, and that's another 8 bucks. Then when you enter this show, you see the grounds full of peddlers selling stuff. They also have to pay for their space. How many do you see? Do you see the dollar signs going into the billions?

What is a registry? — Registration means to record something. Dog registries keep records about each dog that registers with them. The names of the owners, the breed of dog, the date of the dogs births, the color of the dog, who the parents of the dog are and so on. The registry club also gives this dog a number so they can track this dog in their computer system. For your registration fee, you get a beautiful parchment registration certificate to hang on your wall. (If you have any space left).

There are "all breed registries" and there are "single breed registries." There are also "breed specialty registries." They only register certain groups of dogs like the sporting breed registry, or the herding breed registry, the poo registry, the mixed breed registry, the rare breed registry, the coursing dog registry, the Seeing Eye dog registry, the hunting dog registry. If you got it, somebody wants to register it!

This breed stuff is pretty organized like a scout club or like the government system. It all starts at the local level and moves up. This is how some of these large registration clubs came into existence. For instance, I can create a breed of dog and sell the pups. The owners of those pups can come together and form a small club. That club would be called a breed club. That clubs interest is in its own breed. The club can get bigger while more folks get interested in this breed. Folks from further away form their own club in their own town. Soon there's a breed club in every state. One of those clubs in my state can decide to represent all the other clubs in the state I live in. The first club (my club) is now called the mother club. Soon all the states have a club that represents all the other clubs in their state. They are still called a breed club because they are only concerned with their own breed.

Now, say there are ten breeds in the United States that have formed just like the example clubs that I gave you. One of each of those folks goes over to someone's house and they talk about uniting all those ten clubs together, say once a year. "Yea, let's put on an all breed dog show!" Those ten people just formed another club. Because they all live in America, they call this club the American Registration Club. Now each time a new breed club is formed within America, representatives from the big club (A.R.C) go to the little breed clubs and ask them if they want to join. Soon the American Registration Club has over 103 breed clubs that have joined them.

Now, say a group of herding dog owners gets together and they don't want to join the A.R.C because the A.R.C doesn't have any shows that cater to this group of dogs. (Your herding group wants to judge their own dogs on their dog's ability to herd and not on how pretty the dogs are.) Then another group of breeders of gun dogs or hunting dogs gets together and does the same thing. Now you have a hunting dog registration club and a herding dog registration club. Voila! That is why today the world has a whole bunch of registration clubs.

Say you are employed by this multibillion-dollar A.R.C and your job is to find more money to bring into the club because a multibillion dollars is just not enough. Say you go to this herding club and you ask them to join the A.R.C and the majority says, "No." What do you do? Well, you get the minority of folks who want to join and you just go ahead and do the paperwork, register the small group under the large group's name! What do you think happens now? Hehe. The large group of folks with those wonderful herding dogs isn't recognized and the A.R.C just stole their breed's name! The rest of the herding club could be recognized, if they joined. Sounds like the mafia to me! Guess what is even more amazing? It really happened!

I want to stay focused on the new "breed club." The breed club that is all by itself. The mother club. You started the club. You and those special people that love that little Jack Russell or whichever breed you have. You guys know your dogs. You know what you want in the breed. You know how they should act. You are a close net group of Jack Russell Terrier lovers and you get together and put on shows and events just for this little dog. You vote on a gal (or guy) that you believe knows the most about the Jack Russell Terrier. A person that is fair and good. You all know this person well, so you vote him/her to be a judge of the breed. You respect that person and that person knows the importance of a winning dog that will go on to breed. The dog you pick top "Jack" out of all the other "Jacks" will beget puppies. You and your people know the grave circumstances that will come of this.

Now that is the backbone and the truth of a darn good breed of dog! That is where the breed lives and survives! That is where it all happens! Not at a national level, but at the grass roots level. Don't let anybody bully you!

Here is a long list of organizations that will gladly recognize your new breed of dog as "pure of breed" even as a "long lost breed." Why should they care if it is untrue? Some organizations in this list have a few scruples and may only recognize your new breed as a breed if you can show them at least three generations. Ha, what's three generations!

Most of them don't even want proof, you know, like a picture or something. Some registration clubs may put your new breed in a category as rare breed. And all the clubs will give you and your breed rules that must be followed or adapted.

Here is something for you to think about. Say you raise German Shepherd Dogs and you register them with one of these clubs. You have a litter of five pups. You tell the breed registry that you had a litter of twelve pups. They send you twelve registration certificates for each of the twelve pups. Now you go to all the pet shops, kennels, ads in the paper, shelters, etc., and every pup that looks like a shepherd you purchase. Then, you advertise your pups for sale and get 600.00 each for every one of those pups. You have the mother and father on the grounds and perspective buyers can see them! Could be done, no? Buyers beware. Know your breed.

Now there is a flip side to this, if you do register with any of these clubs:

#1 Your breed gets recognized faster by the majority of folks on the planet.

#2 Your breed gets to compete in beauty contests under the registration clubs judges. (Everyone who watched the world skating champions knows judges can be bribed.)

#3 You may make a good bit of money.

#4 You do not have to worry over all that paper work.

List of breed registering clubs

Take a long look at all these registering clubs! Keep looking. I probably even missed some.

World Dog Clubs:
The Federation Cynologique International (FCI) was created on May 22nd, 1911. Their goal was to protect purebred dogs by any means it considered necessary. (I just want to make sure you read that word correctly "protect").

Like all other breed registrations they make even more money with a magazine about their organization. Information on that magazine is available from: Stratego, Muhlenweg 4, 7221 Marz, Austria.

Germany: Kartell Fur Das Deutsch Hundewesen En Und Die Delegirten Dommission.

Austria Osterreichischer Kynologenuerband.

Belgium Societe Royale Saint-Hubert

France Societe Centrale Canine De France

Netherlands Raad van Beheer op Kynologisch Gebied in Nederland.

North America North American Mixed-breed Registry

United Kennel Club (UKC)

American Kennel Club (see United States)

World Kennel Club

National FCI Founding Nation Members:

The Federation Cynologique international (F.C.I.) disappeared during World War I then, in 1921, France and Belgium recreated the international breed registry. Now the Federation Cynologique International is the "World Canine Organization." (Pretty impressive for the country of France don't you think?) This club recognizes over 330 breeds with each breed being the property of a specific country. The owner countries of the breeds write the breed standards. They have their own judges and they put on international shows as well as working trials.

Argentina	Federation Cinologica Argentina
Australia	Australian National Kennel Council
	Canberra Kennel Association
	Royal New South Wales Canine Council LTD
	Victorian Canine Association
	Canine Association of Western Australia, Inc
Austria	Osterreichischer Kynologenuerband
	Austrian Breed Association
Bahrain	Kennel Club of Bahrain
Belgium	Union Royale Cynologique Saint Hubert
Belorussia	Belorussian Cynological Union
Bolivia	Kennel Club Boliviano
Brazil	Confederacao Brasileira de Cinofilia
Bugaria	Federation Cynologique Bulgare Pres de Union des Chasseurs ET des Pecheurs de Bulgarie
Canada	Canadian Kennel Club

Chile	Kennel Club de Chile
Columbia	Associacion Club Canino Colombiano
Costa Rica	Asociacion Canofila Costarricense
Croatia	Hrvaski Kinoboski Savez
Cuba	Federacion Cinologica de Cuba
Cyprus	Cuyprus Kennel Club
Czech	Ceskomoravske Kynologicka Unie
Denmark	Dansk Kennel Klub
Dominican Republic	Federaciaon Canina Dominicana
Ecuador	Associacion Ecoatoriana de Registros Caninos
El Salvador	Asociacion Canofila Salvadoeria
Estonia	Estonian Kennel Union
Finland	Suomen Kennelitto Finska Kermelklubben RV.
France	Societe Centrale Canine
Georgia	Federation Cynologique de Georgie
Germany	Verband fur das Deutsche Hundewesen
Gibraltar	Gibraltar Kennel Club
Greece	Kennel Club of Greece
Guatemala	Associacion Guatemalteca de Criadores de Perros
Honduras	Associacion Canofila de Honduras
Hong Kong	Hong Kong Kennel Club
Hungary	Magyar Ebtenuyeszik Orszagos Egyesulete
Iceland	Hundarrektarfelag Islands
India	Kennel Club of India
Indonesia	All Indonesia Kennel Club
Ireland	Irish Kennel Club
Israel	Israel Kennel Club
Italy	Ente Nazionale Della Cinofilia Italiana
Japan	Japanese Kennel Club Inc
Kazaskan	Union of Cynologists of Kazzkistan
Lativia	Latvijas Kinoloiska Federaciya
Lithuania	Lietuvos Kinologu Draugija
Luxemburg	Union Cynologique Saint Hubert da Grand Duche de Luxembourg
Macedonia	Kennel Association of the Republic of Macedonia
Malaysia	Malaysian Kennel Association
Malta	Malta Kennel Club

Mexico	Federacion Canofila Mexicana a.c.
Moldavia	Uniumea Chinologica din Moldova
Monaco	Societe Canine de Monaco
Morocco	Socaete Centrale Canine Marocaine
Netherlands	Baad van Bebeer op Kynologgisch Gebied in Nederland
New Zealand	New Zealand Kennel Club
Nicaragua	Asociacion canina nicaragtiense
Norway	Norsk Kennel Klub
Panama	Club Canino de Panama
Paraguay	Paraguay Kennel Club
Peru	Kennel Club Peruano
Philippines	Phiippine Canine Club
Poland	Zwiazek Kynologiczny w Polsce
Portugal	Club Portugues de Canicoltura
Puerto Rico	Federacion Canofila de Puerto Rico
Romania	Asociatia Chinologica din Romania
Russia	Russian Kynological Federation
San Marino	Kennel Club San Marino
Scotland	Scottish Kennel Club
Singapore	Singapore Kennel Club
Slovakia	Slovenska Kynologicka Jednota
Slovenia	Kinoloska Zveza Slovenije
South Africa	Kennel Union of Southern Africa
South Korea	Korean Pet Animal Protection Association
Spain	Real Sociedad Central de Fomento de las Razas Caninas en Espana
Sweden	Svenskia Kennelkhibben
Switzerland	Societe Cynologique Suisse
Taiwan	Kennel Club of the Republic of China
Ukraine	Ukrainian Kennel Union
United Kingdom	The Kennel Club British Dog Breeders Council
United States	American Kennel Club
	American Rare Breed Association
	United Kennel Club
	Rare Breed Association
	World Kennel Club
Venezuela	Federacion Camina de Venezuela

Wales   Welsh Kennel Club
Yugoslavia   Jugoslovensky Kinolosky Savez

Do you get the gist of this list? Are you now beginning to understand the vast amount of money that the dog world brings in? And you want me to register my dogs with these organizations! I don't think so!

Breed clubs should keep the money in their own club. Just like the cities should keep the city money in the city. It helps the city!

Why the heck would anyone consider giving money to another organization to take over his or her breed club? Especially if it's Germany, France and Belgium? (That's the F.C.I if you weren't paying attention). The F.C.I registers over 300 different breeds.

Oh yeah, I forgot, the reason folks join a registration club is to be recognized! And to compete in the dog shows. Now put on your business hats and think about this. Why don't we just have our own little club, rule our own little club, bring money into our own little club and judge our own little dogs? After all we are the ones who really know the breed! You don't have to continue this charade. I will give a standing ovation to all those breed clubs who have the brains to keep their breeds out of this mess.

Chapter 4

The Official Breed Standards
Incorporated by the Alsatian Shepalute Club of America @ 2003 and first written in 1978

Trixy Von Der Schwarz

Knowing your Alsatian Shepalutes Standards of the Breed –
The Alsatian Shepalute is a companion dog selectively bred out of the working and herding dog breeds that were recognized as pure of breed by the registry clubs around the world. The Alsatian Shepalutes' pedigrees may be traced back to the very first dogs recognized with those kennel and registry clubs. Knowing something of the background of the dog breeds that have gone into the making this new breed of dog will help the new Alsatian Shepalute owner understand his pup. And then again all Alsatian owners should understand that those breeds that went into the making of this new breed were used only for a particular character that would

enhance the Alsatian Shepalute towards its own completely distinctive identity. That means that in the development of this new breed, pups that exhibited traits of the working dog or herding dogs were not desired. Only certain pups were selected to continue in the breeding that exhibited the desired traits of a family companion dog. Absolutely no hyper ness was tolerated. Therefore all the hyper dogs found homes. They were not used in any of these breedings. All puppies that whined or cried were also not bred. Selective breeding such as this begins to form a different character than that of the dogs that were used in the beginning of the breeding program. The new and desired character and its physical makeup are written down in the following descriptions or standards for this breed.

General appearance — The Alsatian Shepalute is a powerful, strong-willed, well muscled, alert animal. He is well balanced, slightly longer than tall, with a definite impression of masculine or feminine. His look is that of intelligence with a bit of secrecy in his yellow stare. He is a large wolf looking dog that comes in many different colors with the predominate color of coat being a wolf-gray.

Character —- The Alsatian Shepalute is fearless and bold but not hostile, moving in a sleek manner sniffing the air currants. He is self-confident, poised and inquisitive. The Alsatian is never nervous, but with a more solid and laid back temperament of curiosity. The Alsatian Shepalute is a dignified, loyal and devoted family companion dog.

Head — The Head of an Alsatian indicates a high degree of intelligence. It is broad and large sloping slightly from between the yellow eyes down to the deep black nose. This breed resembling the wolf of yester years. The Alsatian Shepalute should have a short coat of hair on the head, face and legs.

The coat should begin to lengthen as it starts down the neck to the shoulders where the hair is the longest.

Muzzle —- His muzzle should be large, his Lips are close fitting and deep black in color with large white teeth. The upper and lower jaws should be broad with his large teeth closing in a scissor bite. The total muzzle should be slightly longer than the head is deep.

Eyes —- His eyes are an almond shape, medium to small, and set obliquely. Yellow or light brown in color with that look that makes him unique, the wolfish stare. A large round eye is a major disqualification.

Ears —- His ears are triangular in shape and slightly rounded at the tips. They are set wide apart on the outside back edges of the skull. The ears are wedge-shaped, erect and small in comparison to the head as well as tipped with black hairs (the blacker the better) to form an outline around the ear. When alerted his ears turn forward. When shamed his ears will turn side ways and laid back.

Forequarters —- The shoulders should be slightly sloping, heavy and muscular without any tendency to looseness of shoulders. The feet are large, heavy, round and compact with well-arched toes. His Leg bones are straight to his pasterns, which are short and strong being bent only slightly. They should have black pads on the bottom of their feet with black toenails. The Black coloring may extend upward from the pads into the leg. The forelegs are heavily boned and set wide apart because of the width of the chest.

Feet —— The feet are very large with compact toes, well-arched pads, thick and tough. He is sure footed even when stalking.

Neck —- The neck is well muscled strong and powerful. It is short in length thick in circumference. The carriage of the head is forward and in line or slightly higher than the shoulders.

Teeth —— His teeth are large and come together in a Scissor bite. Never overshot or undershot.

Coat —- The outer coat is moderately coarse. It is thicker during the winter months. Not too long. Moderately dense, slightly oily and slightly wooly. Thicker fur around the neck. Moderately short to medium along the sides of the body with the length of the coat increasing somewhat around shoulders and neck, down the back and over the rump. The coat is shorter thinner during the summer months. As the undercoat almost entirely sheds out.

The coat becomes thick and woolly again during the winter months. The head, inner ears, fore-face, legs and paws should be covered with short hair. Faults in coat include soft, silky, too long outer coat, too wooly and or curly.

Puppy coats have a thick undercoat with few guard hairs tipped with black. This puppy coat will shed out and the adult coat will come in coarser. All puppies lighten in color as they get older. All puppies have a dark dorsal strip with a black triangle mark over the scent gland area on their tails.

Black muzzles lighten up around the eyes with the nose always staying black. All pups should be born with black noses and dark skin.

Hindquarters —-The Shepalute is broad and powerfully muscled through the thighs. His stifles are moderately bent. His hocks should be set back and wide apart. Parallel when viewed from the rear. The legs of the Shepalute must indicate an unusual strength and tremendous propelling power. Any indication of unsoundness in legs or feet standing or moving is to be considered a serious fault.

Tail —- The tail is set moderately high and reaching to the hocks or a little below. Never sweeping, curling or long. Shorter is more acceptable than longer. The tail should be wide at the root, tapering to the end. The tail should be well furred, but not too long, hanging down when at rest and not curved. Carried high when working or excited yet never curling tightly over the back.

The tail should also never go under the body or between the legs. When sleeping the tail may be curled around the body for warmth.

Color and markings —- All colors are permitted. Timber wolf gray is the most desirable. Noses always stay black and the skin is dark in pigmentation. Ears are outlined in black as well as the tip of the tail. Muzzles can be white or cream.

Size —- No shorter than 24 inches and sometimes getting up to 28 inches from the shoulders to the ground. Weight should appear heavy due to the large bones. Side view movements to evaluate strength of back, reach, balance and top line. The propulsion should come from the hindquarters while the front takes the thrust, balance and coordination. The overall length of the dog is to be slightly longer than tall because of the chest and the tail.

Gait —- The gait denotes power and strength and is very important for the overall look of the dog. The rear legs should have drive, while the forelegs should track smoothly with good reach, but never a high step. In motion, the legs move straightforward. As the dog's speed increases from a walk to a trot, the feet move in under the centerline of the body to maintain balance. The back feet pass the front paw marks left in the dirt. The fast walk is smooth and the top line hardly moves, but glides along with the dog. The dog's head should be in line with his body or slightly higher, but never jetting and pulling the owner with unleashed energy. Cautious, yet never nervous nor afraid.

Judging —- Important in judging Shepalutes is their resemblance to the large Dire wolf. A large heavy boned animal with a large head and large feet. A powerfully built dog with sound legs, good feet, deep chest, powerful shoulders, steady, balanced and sleeking. A quiet dog that is alert to its surroundings and quiet in its stalking. Alsatian Shepalutes should not be hyper. They are a calm dog, their eyes not missing anything. Their ears aware of the smallest sounds. Never nervous, but may be cautious. Loud noises do not make this dog bolt. The eyes should appear to look right through you.

Absolutely no part of these animals should be altered including dewclaws being removed, haircuts, nose blackened or coat colored.

In judging Shepalutes for any kinds of shows, testing of the dogs must be preformed as follows:

1. Starting pistol shots shall be fired 3 times. Dogs must not bolt or run away. Dogs may crouch slightly. The dog's body may express concern. Dogs may give a throaty growl. A competing show dog may not bark.

2. A stranger must pass each dog two feet away from the dog's front view. The stranger must not look at the dog's face. Dogs may lean forward to sniff the air currant, wag their tail or look for owner/handler approval.

3. The judge must not startle competing dogs. The judge must not threaten the dog in any way. Dogs may stare the judge down. Owner/handlers must be in control of their dogs or be disqualified. If a Judge feels threatened by any dog at any time the judge may decide to disqualify that dog. Dogs must accept an unthreatening stranger without barking, growling or biting.

4. All shows must be video taped for fairness and in case of conflicts.

Serious faults —- Shyness, Small heads, round or brown eyes, pink nose, noisy, splay footedness, weak legs, cow hocks, bad pasterns, straight shoulders, lightness of bone, over-angulated. Legs should move in straight lines coming and going. Dogs may gait sideways and not be faulted. Never nervous or shy.

Scale of points

General character and intelligence	20
Face and muzzle	15
Eyes	11
Forelegs and feet	10
Back, loins and flanks	10
Hind legs and feet	10
Height and substance	5
Skull	5
Ear	5
Tail	4
Chest and ribs	3
Coat and color	2
Total points	100

Chapter 5

The Alsatian Shepalute Club

Organization of the A.S.C.

Membership—- The most important element in the A.S.C structure is its membership. I cannot stress this fact enough. Breed clubs who have joined large registration clubs have sacrificed their dogs to the world of fashion and breeding's with winning titles where the lives of humans and dogs alike are put on the line. Greed, jealousy, hate and discontent of fellowmen is hard enough in today's society without joining groups that bring such actions to the surface.

Jealous breeders poison other breeder's dogs just to reach the top! Keeping this breed club solely run by its members and the owners of the Alsatians will guarantee that this breed will not rise to the level where such hatred amongst men thrives.

Management of the club is exerted in two ways as follows:
1. All members elect the officers and board of director's.
2. The basis for the operation is its by-laws, which are approved by the chief executive officer.

Board of directors—- The board consists of three officers, a president, vice president, and the directors. The members elect the officers at large. The board of director's are responsible to the membership for the management of affairs. These are not salaried positions.

Mrs. Schwarz—- Mrs. Schwarz is the inventor of the breed legally. She is the sole judge of all Alsatian Shepalutes in comparison to the standards of the breed. She also is the sole judge of the quality of the offspring of all Alsatian Shepalutes until she passes on this position to a competent member of the club. Her function is in her ability to read Alsatian Shepalutes and to know which dogs should breed and which should not, in order to produce the offspring necessary to uphold the highest standards of this breed. As the quality assurance inspector so to speak, any dogs (bitches, puppies included) who shall carry the name "Alsatian Shepalutes" must be approved by her standards and will thus be given the official certificate of true quality of breeding until this breed is strongly set in its genetic structure. At such time Judges shall be appointed by the members.

Administration of the club

President—-The president is the chairman of all meetings of the members and the board of directors and is an ex-officio member of all committees and councils. The president is the chief executive officer of the club and chairman of the board and has the same rights and privileges as usually vested in the office of president of similar organizations.

The function of the chief executive officer is to develop long-term goals for the club over a ten-year period and in consultation with the staff and the steering committee, develop the necessary plans to attain those goals. The chief executive officer promotes and encourages the aims and objects of the club. This is not a salaried position.

Secretary-treasurer—-The Secretary-treasurer is the chief administrative officer who has general powers and duties of supervision and management and is responsible to the board for the day to day operations of the A.S.C. office, including the necessary financial and budgeting controls for the efficient management of the club.

The staff is divided into: Three sections as follow:
1. Administration
2. Registration Division
3. Shows and Trials Division

Administration Staff — The secretary-treasurer directly manages the administrative staff. This area is responsible for the necessary preparations for the annual general meeting and all meetings of the board. The secretary-treasurer is the secretary of the board and with the help of the heads of registration and shows and trials division implements virtually all decisions of the board. There are however, certain decisions of the board that apply directly to councils and committees rather than to the administrative staff.

Registration Division—-The registrar is responsible to the secretary-treasurer for the handling and processing of all applications and documents relating to registration records for purebred dogs. The duties of this division also include the following:

1. Working with the registration committee on all files requiring the consideration of that committee and is responsible for the implementation of its decisions which are reported to all members of the board.

2. Processing all applications for registration of kennel names. Maintaining master records of the membership. These records are updated yearly and address labels are provided for the mailing list.

Shows and Trials division— This division is responsible to the secretary-treasurer for the following:

1. Operation of shows and trials and planning all work having to do with all that precedes and follows a show or trial.

2. Show and trial results and the final preparation of the material appearing in the A.S.C. Official section.

3. The division supervisor's work closely with the councils and committees included with shows and trials.

Membership
Membership of the A.S.C. indicates positive support for the promotion of high quality dogs.

Member's responsibilities:

To act in general for the welfare of the breed.

To be a responsible dog owner.

To know the breed and breed standards.

The Alsatian Shepalute club members and dog owners meet once a year in February over the Internet. For more information, Pamphlet's and booklets about these dogs and/or the club and how you may participate, please e-mail: Shepalutes@aol.com

Or write to: Alsatian Shepalutes
2080 Antelope Rd. Suite 334
White City, OR 97503

Chapter 6

The Alsatian Shepalute Registry

The Alsatian Shepalute Registry has been in existence since 1988, then under the name of the North American Shepalute club.

The Alsatian Shepalute Registry only registers Alsatian Shepalutes and is concerned solely with the validity of the registration information submitted and certified by the applicant thus keeping our full efforts directly on and towards this breed. We do feel that only such a registry can offer complete alliance to the breed.

Our objective is to:

Adopt and enforce uniform rules and regulations for persons interested in exhibiting, running, breeding, registering, purchasing and selling Alsatian Shepalutes.

To detect, prevent and punish frauds in connection therewith.

To protect the interest of its members in which rules and regulations have been designed to carry out these objectives.

We recognize and work together with the Alsatian Shepalute Club towards the betterment of the breed and open a sincere invitation to any individual who can produce proper and legal proof of the purity of their dog as being that of the Alsatian Shepalute.

We look forward to your membership. Please do not hesitate to call on us if you have any questions or concerns.

Jennifer L. A. Kingsley
President

Contact us at our e-mail address:

Shepalutes@aol.com.

Our registry service is available to those clients who own an Alsatian Shepalute of recognized pure lineage. This is a unique identification system, which compiles a complete profile on each and every pet registered for its entire lifetime when the buyer of an Alsatian Shepalute registers his animal with our club. The names of the sire and dam in addition to the dog itself are permanently recorded in our files. The dogs are assigned registration numbers, which in turn are forwarded to the breeder. These assigned numbers are permanent and always thereafter recognized by this club.

Rules Applying to Registration

1. The purpose of the A.S.R is to certify the pedigree of purebred dogs registered with the club.

2. The A.S.R is concerned solely with the validity of registration information submitted and certified by the applicant.

3. Complaints must be made in writing.

4. The A.S.R does not arbitrate complaints concerning quality, size, conformation and disposition. (These are club matters and responsibilities).

5. The complainant must pursue all litigation concerning such.

In the interest of preserving the purity and integrity of the breed the A.S.R strongly suggests that applicants voice their opinions at club meetings. Our jurisdiction is limited to the practices to be followed in order to ensure proper identification of dogs registered with the A.S.R. The fact that an animal is registered does not, of itself, guarantee its quality or health.

The A.S.R was formed principally for the protection and advancement of the ALSATIAN SHEPALUTE.

Please be sure to read the applications over carefully before filling them out so that you understand what information will be needed.

The sire and dam must be registered before you can register the litter. The owner of the dam at the time of breeding may submit the enclosed applications.

A photograph of your dog must accompany the application and will not be returned to the sender.

General Information of Single Registration

The A.S.R certificate is distinctive evidence of a purebred dog, thus identifying it as such. It also indicates that the animal's present owner is on record as well as the owners of the sire and dam and the names of the dog's ancestors.

1. All applications for single registration must be made on A.S.R application forms, which will be furnished free and will not be returned to applicant.

2. If the applicant is unable to fill out the blank application complete in every detail, this shall constitute doubtful breeding and the dog will not be accepted for registration.

3. The A.S.R offices are a separate and distinct registration office and have no connection with any other registration office whatsoever.

4. The information of this application is the sole responsibility of the individual signing it.

The A.S.R reserves the right to correct or revoke any registration certificate issued. Any misrepresentation of this application is cause for cancellation and may result in loss of all A.S.R privileges for those individuals who violate the integrity of this application. We reserve the right to change any and all rules, fees, etc. without notice, anytime conditions warrant it.

Application for an Alsatian Shepalute Single Dog Registry Registration fee:
The fee for single registration is $10.00
Mail to: Alsatian Shepalute Registry
2080 Antelope Rd. Suite 334
White City, OR 97503

Office use only {   }
Date Received   {   }   attach photos here
Fee {   }

Date of birth _____ Color _____

Tattoo or brand _____

Print dog's name __ __ __ __ __ __ __ __ __ __ __ __ __

Owner of sire at time of breeding _____

Owner of dam at time of breeding _____

Sire _____ number _____

Dam _____ number _____

(If the sire and/or dam are registered give their complete names and registration numbers.)

This is to certify that this pedigree is correct to the best of my knowledge and belief.

Sign name _____

Print owner's name_____

Street _____ _____

City _____State _____ zip _____

Telephone number _____

E-mail _____

Lois Denny

Application for Alsatian Shepalute Registration for a Whole Litter
Fee for registering a litter $25.00
Mail to: The Alsatian Shepalute Registry
2080 Antelope Rd. Suite 334
White City, OR 97503

Name of sire _____   reg. No. _____
ATTACH PHOTO

Name of dam _____   reg. No. _____
ATTACH PHOTO

Owner of sire on date of mating fills out this section:
I certify the above dam was mated to the above sire on _____

Signature of owner of sire _____

Address _____

City, state _____

Owner of dam on birth of litter fills out this section:

I certify that I was the owner of the above dam on the date of the litter. That the number of puppies in the litter as shown below are living and that this pedigree is to the best of my knowledge and belief true.

Signature of owner of dam _____

Print name of owner of dam _____

Address _____

City_____ State _____

Month of birth _____ day _____ year _____

Male's _____ females _____

Yes, put my name on the A.S.R breeders list! I authorize the A.S.R to give out my name, address and telephone number to prospective buyers looking for pups and stud service in my area. Include 10.00

Signature_____

E-mail_____ Telephone _____

Application for Registration for a Whole Litter

To register a litter, fill out the litter application. Make certain that the registration numbers of the sire and dam are correct and with the correct ownership. You can make certain by checking the name and address on the front of both sire and dam's registration certificates against the individual who has represented themselves as the owner.

We look forward to receiving your single dog and litter applications. If you have any questions, please do not hesitate to contact us: Shepalutes@aol.com

Forms available from the A.S.R for a slight fee:

1. Registration Certificate (duplicate copy)
2. Registration Fee for a Single Dog
3 Registration Application of a Litter
4. Lost Registration Papers (duplicate copy)
5. Name Change of Dog
6. Transfer of Ownership
7. Certified Pedigrees up to 20 generations
8. The Alsatian Shepalute Breeders Book
9. Breeding Practices and Principles
10. The Alsatian Shepalute Book

Corporate By-laws

1. The Alsatian Shepalute registry will work for the protection and advancement of the pure-bred Alsatian Shepalute.

2. Complaints of false information must be submitted in writing, and will get 45 days time to investigate and respond to such complaint.

3. Application for registration must be completed with all requested information and evidence of eligibility for registration must be submitted with such.

4. Application for a single dog must be accompanied with four pictures of the said dog. Front view, two side views and a rear view of said dog in

a standing position. The photos will remain the property of the A.S.R and will be permanently placed in your member's file.

5. The sire and dam must be registered before a member will be allowed to register a litter. All requirements of the application must be met before a registration certificate of membership can be issued.

6. Litter registration and application for litter registration must be completed in full, making certain all registration numbers of the sire and dam. Applicant must check the name and address on the front of both the sire and dam's registration certificate against that of the individual who has represented themselves as the owner. All applications must be received within six months of the whelping date. A picture of the whole litter must be accompanied plus a picture of each pup. The pictures will remain the property of the A.S.R and will be permanently placed in your member's file.

7. Breeders must keep records of the puppies sold and register them in a book providing the sire and dam's name and registration numbers, the whelping date and the registration numbers of each puppy that has been sold.

8. Transfer of dog. The back of the issued certificate of registration must be completed to transfer a dog into another name and ownership along with the appropriate fee.

9. Most owners will change the dog's name after you have sold them. Breeders sometimes use their own surnames and give them a male or female name, or if you prefer, you may use the kennel name where your dog is registered.

10. Inspections. The A.S.R authorized representative has the right to inspect the records and practices and to examine any dog registered with the A.S.R

11. Penalties. The A.S.R may refuse any dog or litter or to record the transfer of any dog for sole reason that the records do not support the application required.

12. All registration records are kept confidential by the A.S.R and any request for file information must be submitted in writing and signed by a member who is the owner of record.

13. The A.S.R reserves the right to change any and all rules, fees, etc. without notice anytime conditions warrant such changes.

14. The A.S.R stands firmly behind the constitutional rights of all individuals.

Chapter 7

The Alsatian Shepalute Puppy

The "V" litter 2003

Selecting Your New Alsatian Shepalute Puppy

A lot of puppy books start out by explaining to the soon to be owner that the whole household should agree and be in on the purchase of a new pup. If that were the case, I would have never had so many dogs in my lifetime.

My mom always picked out the family's dogs. I got my own dog when I was about 13. Then I got my second and third dogs and then I had to get my own place!

Since then I have had over two thousand dogs in my lifetime, perhaps more, and I have loved every one of them!

Of course, you should know that owning a pup means more than just 'hanging' around with your pal, unless you are fortunate to live in the mountains, hilltops, and farmlands! Even then the most important thing I can think of for you to realize is, to know the laws where you live concerning dogs. It's important for your piece of mind and so that you don't accidentally upset your neighbor. It is also important in that no one will be able to trample on your constitutional rights in owning a pet. The law states that not knowing the law is no excuse. But that is for another chapter, so make sure you read it!

Know also that a pup needs its shots so that he doesn't get sick. It is a hard fact of life when you're first pup dies of a sickness because you didn't realize you needed to get the pup its shots. (When I was young I could never understand why any of God's creatures needed shots when they all did so well by themselves.) Make sure you read those boring chapters about the health of your new pup. I'll try to give you all the facts you need to know from a laymen's point of view.

So the first thing you should look for when selecting your new Shepalute puppy is not the looks and beauty of the pup or whether it has a patch of white on its toes! You better check out the pup's disposition! But how can you tell about the pup's character when the pup is so young?

Here is some food for thought: As you look at the litter of pups think to yourself, what will that pup be like as it matures? Make sure you check out the father and mother of the pup and if you can, the grandparents. It has been my experience that all pups resemble their grandparents rather than their immediate upline.

The Sire and the Dam

How do the parents act? How old are they? If the pup's parents are three years old or younger, then you must realize that they are still pups themselves. If the pup's parents are ten years old or older, they may still think they are pups but in reality they are slowing down. (Just something to consider.)

If the parents (the sire and the dam) are well trained, it will be hard for you to tell how your pup will turn out, unless you have the uncanny ability to be able to put aside the human conditioning and/or environmental conditioning that has influenced those dogs since their birth. Human and environmental conditionings always play a big role in the animal you see before you. If the parents have not been trained then you will be able to quickly tell how your pup will be when it matures. The untrained animal is the true character of the dog and it is the only way anyone can be absolutely sure about a new pup when sizing up the parents of that pup. (Keeping in mind the environmental conditioning that has influenced that adult dog.)

I am a witness to the judgments of many folks who were out looking for a pup and the pup's sire and dam were well trained. The public actually does think that this new pup will turn out just like the parents! So well behaved! That just doesn't happen!

Of course you will also influence your chosen pup or 'mold it' out of the parent stock into the companion dog that will fit into your lifestyle. Remember that the time you put into a pup from the moment you take him home until he is at least ten or eleven months old is the most important and influential time of that pup's life and of your sanity. Figure it this way, the more time you spend with your dog in his first nine months is equal to the obedience your pup will give you in return. I am a stay at home dog mother, so to speak, and I go for walks with my pups for my own health as well as the pup's trainings. I get about seven hundred fifty hours with my pup before he turns nine months old.

You should not have much trouble with the Alsatian Shepalute in those troubling adolescent years. Your Alsatian pup should not be a hyper dog. Puppies will be puppies though and you must take that into consideration. They do get bored.

The Chosen Pup

When looking for that special Alsatian pup make sure you handle all the pups in the litter and when you pick that special pup up, ask yourself these questions. Does the puppy squirm and cry and bite to get down? If it does it cannot be an Alsatian Shepalute!

Does it grab a hold of your neck with both front legs? You may think this is a sign of love, but it is not. It is a sign of fear of falling down or uncertainty.

If a pup shows that clutching sign, he may be a bit insecure and should be trained accordingly. Proper training can and will give the dog a more stable personality. Just note that your pup was that way and if you intend to breed him, don't! This type of pup will bond quickly and will heel automatically.

The perfect personality in an Alsatian Shepalute is a pup that is comfortable with you picking him up and wants to lick your face and wags his tail. Even one who just accepts it and is loose and comfortable is fine. As a matter of fact, I would choose that laid back pup myself.

If the dogs you go to see are kennel dogs they are pretty close to a pack formation and they are in their element both dominant and recessive. Ask to see the dog's parents outside the kennel and on a leash so you can see how your pup might be as an adult. If the leash is dropped what will the dog do? Does the dog run off? Does he bark and run and jump around or does he settle down when talked to? He may just be excited so take your time and just watch. Sit down and observe the dogs, all of them. How does he approach you or does he? He should be unconcerned with you. He may sniff the air or sniff your jeans to find out who you are. You may kneel down or sit and he will come off his guard of you. If he is growling at you or barking at you it may be his training. Of course you do not want to see the parents then. That doesn't mean the pups will be that way. A barking dog is a shy dog. He is frightened and will usually run away. Remember I am talking about a raw dog, one that is not trained. My dogs are not allowed to bark at you when I am around. And they usually don't. They are comfortable in who they are and they are not afraid of you. Don't forget you are a stranger. If I let my dogs out of the kennel they would pay you no mind. If you sat and called them over they would come easily to be petted, but their sole interest would be in me and what it was we were going to do. The puppies on the other hand love to get out of the kennels and run busily around getting into trouble, sniffing things and checking stuff out. They would be excited for about ten minutes then they would sit in your lap. Where are the hot dogs? Shepalutes are not hyper when petted. They will sit there quietly. They will not squirm, cry, bite, or jump all over you. That is the quality of a dog that is easy to train and easy to control. That is a dog that will fit nicely into any family situation.

The next thing you do is to ask for a business card and the names and prices of the pups. Take notes, and then go home. Think about it. Go visit other breeds and other pups. Pick them up and feel the difference. Read the standards of the breed and be realistic about the size and weight of the adult dogs as they would fit into your family's situation. Get business cards from all the places you visit and make notes, then go home and think about it!

Please whatever you do, do not judge the pups from the attitude of the humans or the environment where the pups were brought up. Nor judge the litter by the house keeping of the owners or their hospitality! Judge the pups by the parents and grandparents and the information the breeder has. Many a good pup may be overlooked, as a pup's total being is not that of the breeder or the breeder's beliefs or lifestyles. If the breeder is a farmer and the pups are in the stable muddy, fat and lively, it is ok. The most important fact is that the owners know about their animals. If they do not know the grandparents or great grandparents of the pups, and the parents of the pups are not visible, run do not walk to your car. You do not need to get a business card. Do not buy that pup or even take it free!

Male or female?

With all the dogs that have gone through my hands as a Groomer, trainer and breeder, I feel that I can tell you with certainty, that the best dog in the world is a neutered male! That is based on the lifestyles of the everyday families of today with most families living in the city. Let's examine both sexes:

Females: They do not roam the neighborhood as a male would; they return home and don't go so far. They do mark there territory but in a different way than a male. They mark the territory to invite male dogs over and to spread the news that they are ready to breed when they come into season. They are the most docile of their breed in their puppy stage until… they come in heat, or somebody takes what's theirs. They are possessive and moody.

What if you spay them? #1. Half of their insides have to come out and God put all that stuff in there for a reason. I don't care how good a veterinarian your vet. is! I believe it is harmful and wrong to put a bitch through that. I will state that of the thousands of dogs that came through my shop once a year that 90% of the spayed females were touch sensitive on their bellies. In my opinion, it was because part of them had been removed. It is what I have seen. Some females who had been operated on twice had to wear muzzles or take sedatives to get through the grooming process. Yes, I said operated on twice! Why you ask? Because these were pound puppies that no one bothered to transfer records on. Being spayed twice dramatically changed their behaviors to where no one could pick them up or approach them.

Let's go on. Females make better guard dogs. They call them bitches for a reason.

Female Shepalutes come into there first heat cycle at around 12 months of age. Female Shepalutes will have a strong desire to mark their territory and to round up and tease the males just as female dogs do of all breeds. Female Alsatian Shepalutes coming into heat will fight off any dogs coming into their territory male or female, as this is the place where they will dig their dens and protect their litters. They are very protective of their territory. Let's say you do spay your female Alsatian. The personality of your dog will change a bit depending on the age the dog was when it came under the knife. If you spay a female at around five months of age, the personality change will be slight. The dog will never be a bitch. She will not be dog aggressive. She will be friendlier. If you spay late in life your female dog will settle down within a couple of months. She won't have a desire to fight off other dogs any longer. She won't wander at all. She will still understand marking territory but she will be more like that neutered male only with 1/2 of her insides gone.

Males: Male dogs lift their legs. Male dogs mark their territory to let all dogs know that they have arrived and that this is their turf! They have male parts that do unsheathe themselves when they get excited. They also lick their private parts more than a female does unless the female is in heat.

Male dogs have a strong natural desire to roam great distances to find the right females to breed with and they do not care where they have to go to find one. They will dig, jump or bust out of their yards to get to a female down the street. I once had a female Rottwieler who when in heat broke down a six-foot wooden gate to get out to tease my stud Alsatian Shepalute.

At around fourteen months of age and in the spring season, males may become restless. They may whine or howl. They are masculine and larger than female counterparts of the breed. They look more mature, poised and alert. Male dogs can get more aggressive with other male dogs than with females only because males won't fight females most of the time because they want to be friends when the time is right. But know that a female dog is more aggressive to strange dogs than a male dog is. The male's just let the females run them for a while, especially if the male dog is not neutered as it is part of his make-up. A male dog is curious and wants the strange dog to check him out or come over so he can see if the stranger can be mounted. A female dog doesn't care about that, unless she's in heat. She doesn't really care for the strange dog and will act tough. If confronted she may turn and run, unless she is on her own turf. The strongest and meanest male wins. That's just the way it is, but here's the secret. Neuter a male pup before his testosterone kicks in and he never knows he is a male! Doesn't care either. His private parts stay in more and he doesn't have to lick himself as much.

A neutered male is good looking and sweet. The real secret is the timing of the neutering! Neutered males don't roam as un-neutered males do. Some neutered males don't roam at all. Neutered males do settle down more than if they weren't neutered. A neutered male bonds closer to a human family than an un-neutered male. The un-neutered male longs for that call of the wild, the neutered male doesn't ever hear it.

If you have a premonition that your male pup will bite strangers and you do not wish for that character, neutering an Alsatian Shepalute early will absolutely slow that characteristic down. Mellow the dog out and make him friendlier. I have seen it. They don't care to leave the family or to pick a fight.

I personally like that stallion like animal. In a large dog, like a Shepalute, I myself would wait till about two years old before I would neuter him. Now if he is macho or on the mean side, I'd neuter him at eleven months, but know that I can handle an abusive tough dog. I also live in the country.

Neutering keeps them mentally in a puppy stage, so to speak. They haven't even started lifting their leg up to urinate until the age of eleven months or so. (Remember, I am talking about Alsatian Shepalutes. Little breeds come into adulthood much sooner. I am a witness to a male Chihuahua breeding a female at 5 months of age.)

Perhaps I am a little prejudice. Male or female, it is ultimately up to you. I must include in this paragraph that as of this writing I have never neutered any of my dogs, I do not believe in it as a religious preference. I give you unprejudiced information so that you can make an educated choice.

The Breeder

The most important thing I can tell you about choosing your next pup is to know the breeder! Listen to them. Do they know their dogs? Here's a great question to ask! "How many generations have you bred?" That question alone tells me if the owners know anything at all about how to breed the dogs they have. Here is another question. "Do you have pictures of the parents and the grandparents?"

When prospective owners come to me to purchase a pup, I can tell right away if they know anything about the greatness of the pups they are observing by the questions they ask. Each individual that comes for a pup is unique in personality, as is each pup in the litter. If the pup is to be a family pet the pup chosen must fit into the family as a whole. The questions I ask perspective buyers would be:

1. Who would be mostly responsible for this new pup's training and up bringing? Whose dog would it be?

2. What does the family do as a whole?

3. If you had an adult dog of this breed what would you require of him? What would you use the dog for?

These types of questions help me to select a pup from the litter that would fit that family's lifestyle. I know each pup and there personalities and I can best choose a pup that will fit into the new family. Most breeders can do that.

How are their records kept? Ask to see the paper work on the pups or parents. Do they have any? It never ceases to amaze me that some folks do not care what the parents act like let alone the grandparents. No wonder so many folks who go looking for a pup for the family end up taking those dogs to the pound!

Lois Denny

I have thirteen generations of picture albums, photos and notes of every single Shepalute I have ever bred. I even have scrapbooks of all the generations further back into the pedigrees of the pure-bred dogs I have bred with. All that is proof that your breeder knows what she or he is doing and that they actually bred the dogs they say they have.

Folks do come over who aren't even interested in the lineage. They aren't even concerned with a pedigree. The pup itself can tell you a lot, I agree, but a new owner must know how their pup will be when grown. If they don't understand that the pedigree and parents of the pup determine that pup's disposition, then so be it. Those new puppy owners do not understand the dog world and may end up not liking the pup they chose just because they did not understand genetics. Then again you may just have that uncanny ability to pick a great pup, then you also have a gift.

Know your pup's lineage first hand. Do not rely on a pedigree alone. That pedigree can't tell you if the parents limped or barked too much or had bad skin problems. And if the breeder tells you some faults of the parents or lineage, be sure that this is a breeder that is telling you the truth for there is no perfect dog when striving towards the breed standards. A good breeder's motto is to "always breed better than the last." Each pup will have its own unique faults to a good breeder. That doesn't mean the pup is faulty!

K-9 Intelligence

This is a misconstrued word. Especially when dealing with animals! What one person considers intelligence in a dog may not be so for another. Somebody once put together an intelligence test for a number of different breeds of dogs. How stupid is that? It received a write up in the newspaper and was on national television!

All you breed lovers, who lost in this intelligence test, don't fret! Each breed of dog has its own type of intelligence depending on what the dog was bred for, and the person who designed that infamous intelligence test just proved to real dog breeders that he didn't know what he was talking about. If you are going to make an intelligence test for dogs it must only be applied within its own breed to hold any type of validity and there must be a clear and defined definition as to what k9 intelligence is.

As a breeder of the Alsatian Shepalutes, I judge intelligence in our breed as a dog that pays attention. A very young pup that understands body language has to be a calm and attentive pup. It can be no other way. That pup is quick to learn and easy to train. In my opinion, body language is the universal language for all species. (Ask for my book "Training the Alsatian Shepalute") All Shepalutes understand body language because they have been bred with the genetic characteristics to pay attention. In other words, they are not hyper. They do not have the hyper genes necessary for survival in the wild. Is that hyper behavior, intelligence? If you live in the wild it is! But when you become a domesticated animal and are required to live in a human's world, the intelligence level of that animal changes to fit into its new environment. Intelligence is rooted in character and is inherited. If you wish to know more on this K-9 intelligence stuff refer to Chapter 2.

Hopefully, you now know what to look for in a pup. What is on the inside is what counts, unless of course you want both! There is absolutely nothing wrong with that! What looks good to you, though, may not look good to someone else. I myself want both!

Bojik Berenty six months old

Cody, Topanga and the "T" litter pups 2002

Bringing Your Pup Home

When you bring your puppy home he/she will be missing the pack. It will cry and howl when left alone. I suggest you bring your pup home on a Friday so that you will be able to give this new pup all the attention you can for at least three days. Get him use to your routine early. Pretend to go to work at your normal hour and peek around the corner, plan it all out. Does he have enough room? Can he get into anything while you are away? Does he have something to keep him occupied? Can you come home at lunch, get off early, and bring him with you?

I suggest you also provide a crate for the new pup and put it next to the bed, close enough for your hand to tap the crate so the pup will know you are there. Please be patient with your new puppy. Shepalutes are companion dogs and do not like to be alone. Anything your pup does in its crate, playpen, or exercise pen is ok. Deal with it, but do not punish the pup. If he potties in his crate blame yourself not the dog. The puppy had no choice. You were not paying attention. Your pup does not want to soil his bed.

Give your pup something to do or to think about. If your pup is whining and crying and carrying on for no reason (because you left him for a second, or just because he is in the crate) then put the crate in a separate room and let him carry on. The reason I say that is so the whining and crying doesn't upset you. If you give in to the whining your pup will win and whining will be a way to get you to come to him. This learned behavior gets worse, never better and you end up with a whining dog that your neighbors will not tolerate. I once knew a man that went down the street and shot a Doberman pup just because it was whining too much.

Do not go to him under any circumstances when he is whining or he will think his crying made you come to him. The trick is to pay attention to him when he is quiet and good. Two minutes after he stops messing around, go to him. Talk to him, pet him, give him a treat, and then go outside the door. Get back to him before he starts up again. Repeat this until he realizes that you have not left him for good.

Here is an orderly list of things I would want if I were going to purchase a pup to bring home as a family pet:

Dog food.
News papers.
Vari kennel.
Puppy shots.
Worm medicine.
Choke chain.

15 ft. Leash.
Large stainless steel food dish.
Water lick or steel water bucket, or both.
Curry comb.
Toenail clippers.
Walking shoes and a pair of old jeans.
Work out coat or jacket.
Cloth training bag to carry his bumpers, leashes, etc.

Don't forget this list is mine not yours. I'm an animal trainer who feels all animals must obey commands, signals and words even if we do live on a farm. I love my dogs too much not to have that communication between us. I know that not training my dog may cost him his life.

Chapter 8

Nutrition

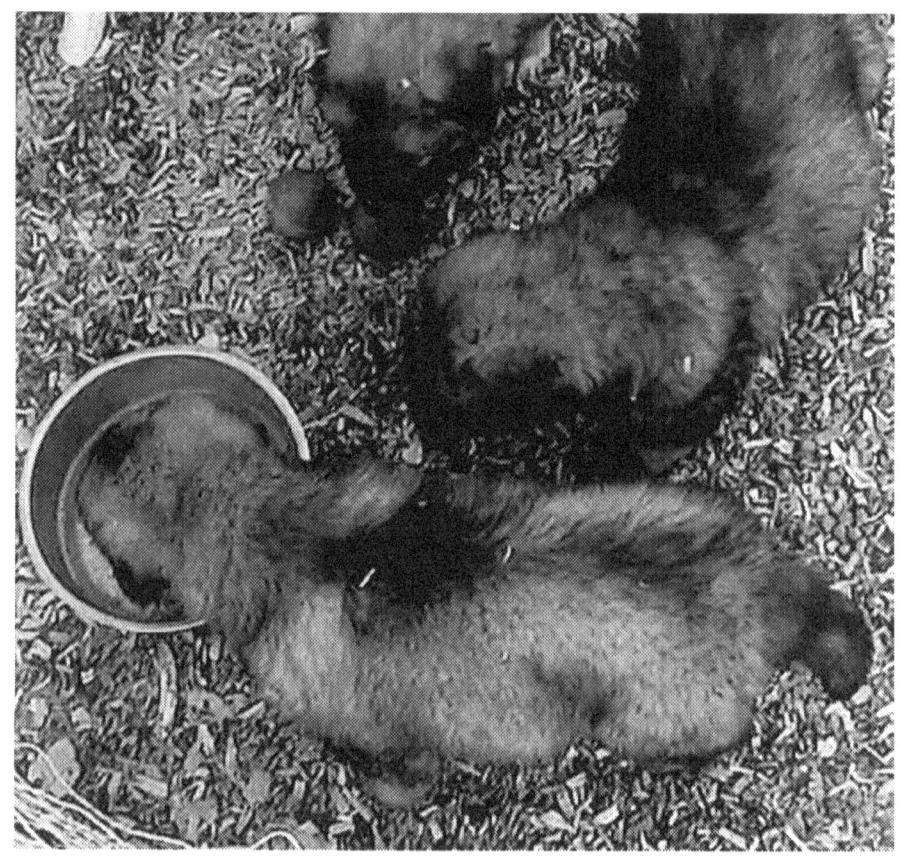

"S" litter 2002

Basic foods and supplements

Protein: meat, dairy products, eggs, soybeans.

Fat: butter, cream, oils, fatty meat, milk, cream, cheese.

Carbohydrates: cereals, vegetables, and confectionery syrups. Honey.

Vitamin A: greens, eggs, milk

Thiamine: legumes, whole grains, eggs, muscle meats, organ meats, milk, yeast.

Riboflavin: green leaves, milk, liver, egg yolk, wheat germ, yeast, beef, chicken.

Niacin: milk, lean meats, liver, yeast

Vitamin D: fish and fish liver oils, eggs, fortified milk.

Iron, Calcium, And Phosphorus: milk and milk products, eggs, soybeans, bone marrow, blood, liver, oatmeal.

Proteins: build body tissue and contain amino acids.

Carbohydrates: give a body fuel for growth and energy

Fat: produces heat, which in turn becomes energy

Vitamins and minerals: act as regulators of the cells activities.

Domesticated animals rely on humans as they cannot live off the land in today's society. We have also bred the domesticated animals to require humans to take care of them, as they would not be able to provide for themselves in what I call the real world. In order for your dog to get everything he needs in the way of nutrition we must combine the foods necessary for his everyday health. Kibbles are fillers to which must be added fat, milk, broths, and meat. When dog food companies bake or cook their ingredients that make up the kibble most of the nutrients are killed. If the dog food companies used the correct amounts of meats and fats and oils, the dog kibble would go rancid.

Not enough has been said about fat other than it is bad. On the contrary, it is what I consider the most important food in an animal's diet.

Fat should be introduced into the dog's diet in its pure form. Proteins and carbohydrates are converted into fat by the body. Fat also causes the dog to retain his food longer in the stomach. It stores vitamins A, D, E and K and lessens the bulk necessary to be fed at each meal. Fat can be melted and poured over the meal.

The animal protein factor in my opinion should be raised. I believe the canines live mostly off fat and proteins and that not enough protein is in the kibbles we feed our dogs. Dogs are meat eaters, so it is important to feed Alsatian Shepalutes meat. Please read the ingredients of everything you feed them. I feed only foods where the ingredients listed on the bag begin with a "meat product" such as: beef, chicken or lamb.

Once I was buying my dog's food (kibbles) and I had to explain the reason why the first ingredient had to be "meat" to the store employee. He had no idea that the first ingredient listed on the bag of any product declared to consumers that the product that they were buying had more of the first ingredients listed. So, the first ingredient may be 50% of the total food and the second ingredient may be 24% of the total food. However, the second listed ingredient cannot be more than the first listed. The third ingredient is less than the first two and so on.

So if your bag of dog food says corn, corn meal, rice, and then chicken in that order, then your dog's main food source is coming strictly from corn. Percentage wise, the next thing you are feeding your dog is cornmeal and then rice and then chicken and so on. This bag may not tell us how much of each ingredient it has in it, so the percentage of chicken may even be at the lowest extreme level!

Corn is the worst food one can feed a dog! Even wheat is better than corn! Corn is not well digested as you may have noticed when you ate corn. Herbivores eat vegetation. Carnivores eat meat. That is their main staple.

The second ingredient I personally would like to see is a 'meat-by-product'. I realize that a meat-by-product is the feathers, legs, beaks, and so on, but don't the wild dogs eat that also? Hasn't hurt them.

Then the next thing you have to realize is all the chemical additives. Try and find a food that hardly has any. Don't feed your dog's sodium nitrite, red dye-40, butylated hydroxyanisole (BHA) or butylated hydroxytoluene (BHT). Also stay away from monosodium glutamate (MSG).

Dog kibbles are filler for my dogs and not really a main staple.

How much to feed?

I fill their bowls. One large bowl per adult dog at least 6 ft apart. (Hehe). As much as they will eat when pups. When they go away from the bowl, I put the leftover kibbles back in the bin. I feed puppies twice a day after each of their lessons. I never give a dish of food to any dog without having them perform something for me. After a pup gets about four or five months old, I will only feed once a morning after our walk and a training session. Training sessions last about five minutes.

My older dogs get fed every other day and there is always a fresh supply of cool clean running water. Wild canines nibble on vegetation when hungry, then about every four days they will kill larger prey animals and they will gorge themselves. I do realize that canines do eat nuts, grass, dirt, rodents and Insects. (Rodents and insects being meat also).

That is the way they are. None of my dogs are thin or overweight. All of them love their food and are willing and receptive to training. Each of my adult dogs eats sixteen cups of kibbles mixed with table scraps every other day.

What I feed my dogs:
Raw bones.
Ground or chunked meats raw or slightly cooked.
Raw organ meats.
Raw veggies if they like them.
Fruits from our trees as they wish. (Grapes, apples, peaches, nuts, pears).
Seeds and nuts including wild birdseeds dropped from the bird feeders.
Fresh eggs with shells still on them.
Table scraps after we eat (mixed into the kibbles).
Bacon grease.

I also add to their 50-gallon trash bin full of kibbles: seafood kelp powder, wheat germ and bone meal (A pinch here and there).

Now I don't go overboard. I feed a small amount of a variety of things.

I have five large water buckets in the kennels. I planted mints of different varieties next to each of them and I pick some and put it in the water buckets.

There is a small stream that goes through the kennels along with tall grasses and a small berry bush. I try to keep it as natural an environment as I can.

I believe that most dogs acquire skin problems along with many different types of ailments that come from poor nutritional habits. The rest of the problems that most dogs have, come from improper breeding and coddling or helping sick dogs get better. In nature the sick and weak die. That is god's breeding program not mine. Don't breed sick or weak pups.

Feeding your puppy

I suggest that the new parents (owners) feed a very young puppy three times a day and an older pup twice a day. Pick up all the food that isn't eaten within five minutes.

12-14 day old pups: evaporated milk, melted butter, an egg, a pinch of brewers yeast. (Three times a day to supplement the mother's feedings).

3 weeks old: 1 tsp. canned chicken dog food is added (three times a day).

4 weeks old: evaporated milk, melted butter, an egg, brewers yeast, can of chicken rich in fat. (Four times a day). I mix this up and put it in the refrigerator. I take out what they will eat and add hot water.

5-6 weeks: puppies are being weaned. Same diet as above only softened kibbles in milk are added. 1 teaspoon of cod liver oil is also added.

7 weeks: pups are completely weaned.

3-5 months: Kibbles are not softened. (Feed three times a day).

5 months-1 year: two large meals are given daily. Table scraps are used sparingly. Just watch the stools and don't spoil the dog. Mix the scraps well into the food.

The rest of his life he will eat once a day or a very large portion once every other day depending on his appetite. Dried milk is added occasionally as well as an egg a day from our chickens as they lay. Alfalfa leaf meal when I can get it. Yeast, sea kelp and bone meal are added to the large trashcans of mixed kibbles.

Feeding ranks next to breeding in the influence it exerts on the growing dog.

Feces 101

Like the Indians and trackers of our past, I know that "one is what one eats" and a lot of the way an animal feels can be seen in the appearance of the feces. If one of my dog's stools is soft, I feed all the cheese, bones, calcium and hard kibbles that puppy wants. If the stool is too hard, I soften the food intake with cooked rice, chicken, fat, grease, or canned dog food. Please check the ingredients of all packaged dog foods. I personally like to give my dog's real food.

If my dog's stool is dark black it tells me that dog has gotten too many minerals in its food. If the stool is chalky then that dog doesn't have enough minerals in its food. If a dog's stool is green then that pup has an infection or has eaten grass. If a pup consistently eats grass, I will check his stool for worms both visually and on a slide under a microscope.

Commercial foods

Dogs do not cook their foods, humans do. Cooking foods kills all enzymes and many of the nutrients. Kibbles are mostly grain, wheat, corn, etc. Again, Shepalutes are meat eaters. Many dogs may be allergic to grain though Shepalutes have never had any of those symptoms that I know of.

When I first started breeding German Shepherds I found a lot of shepherds to be allergic to corn and wheat besides pollen and grass. I did run into some types of food allergies, along with skin and ear problems. In all of the generations of Shepalutes I have bred, I have not seen any of these problems within this breed, but one never knows. For this reason, I ask all Shepalute owners to continually contact me with all symptoms and problems of the smallest degree, so that I may note it in my records.

In my opinion, the best guide to feeding your Shepalutes is to think about what the wild dogs eat. I have seen wild dogs eat the grapes off my grapevines and the fallen apples, rotten or ripe. I have seen nuts of the many shrubs in the Los Padres National Forest in the feces of the coyotes as well as squirrel hairs and hairs of other rodents. Wild animals drink from natural streams and chew on grass and even eat dirt. I don't stop my dogs from doing their thing. I try to understand the vitamin or mineral needs of my pups by observing my animals. When I see them eating the dirt that gophers have dug up, I add calcium and bone meal to their foods.

Myths: (Un-Truths)

1. The digestive system of the modern dog is different from that of his ancestors and therefore must be fed differently.

2. Dogs shouldn't eat bones and raw foods.

3. All dog food should be cooked.

4. The best way to feed a dog is with commercial dog foods.

Chapter 9

General care

Schwarz Kennels 2002

Planning a Kennel Run

In my opinion a housedog or even a cat should be provided with an outside run and house. A place of his own to keep him in the sun and air and protect him from disturbance by children or injury. In this run, your dog is safe from accidents, theft, poisonings or getting out of the yard.

The run should be as large as your property will permit. Twenty by forty feet is a good size for one or two dogs, but if you have more space a longer run is preferable. Length is more important than width.

The best surface that I have found has been straw or chips with sand underneath.

Dirt runs become muddy in the wet season and dusty in the dry season and the dogs waste is hard to pick up.

Something to remember is that dogs usually defecate in the same area and as far away from the food and housing as possible.

Building your run — First remove the grass on the area to be converted into a run. You may want to spray paint the perimeter. Then with a roto-tillier loosen the dirt to a depth of about four inches. Scatter dry cement at the rate of two-thirds of a sack of cement to a square yard of surface and mix in thoroughly with the soil until the mixture has a floury texture. Wear a nose mask to keep the dust out of your nose and safety glasses for your eyes. Adjust your hose to a mist spray and water the surface until the soil-cement mixture will mold under pressure and not crumble. Follow by raking the entire mixture to assure uniform moisture, and level at the same time. Remember to keep a slight slope on all run surfaces so that water can drain off without puddling.

Now you must work quickly, compacting the run with a tamper and then rolling over it with a garden roller. All this must be done within a half-hour or the surface will harden while still uneven. After rolling, the surface should be smooth and even. Mist-spray again, then cover with a coating of damp sawdust or soil for a week, after which the run can be used. Soil-cement is also excellent for paths around or to and from the kennels.

Cleaning your run ———- In removing stools from a run, never rake them together first. This practice tends to spread worm eggs over a greater area. Shovel each stool up separately and deposit it in a container. Your pet store has a rake and shovel pan with a long handle on it that works great! It is lightweight and you don't have to do much bending. For a container, I use those plastic washing machine soap buckets. Those buckets work well for water buckets and when they get old I use them for stool pick up. I put the plastic bags I get at the grocery store inside the bucket and when the bag is full, I tie two knots in the top and put that bag in another bag and deposit it in the garbage.

Having a lot of dogs, I also use a wheel barrel and roll it right into the kennel. Since I use straw or chips I can dump the contents in the pasture and rake it out as smooth as I can. Then, I throw some dirt over it. It is biodegradable and helps improve my ground. I then hose out the wheel barrel and apply a disinfectant. In the winter, we get a lot of rain and some snow so very little can be done about run sanitation. Because of the cold weather, worm eggs and fleas are not present. You can only do just so much for the sake of appearance and to keep down the odors. I purchase a load of straw and in between raindrops I cover the wettest places up front with fresh, clean straw. This keeps my dogs up out of the mud.

Fencing your run —- The ideal fencing would be heavy chain link with metal supporting posts set in cement and erected by experts. Ha! What I had to do was to buy kennel sections that you just put together with clamps. Your local fence company should be able to help you. They cost about 80.00 a section. I found that the 10-foot long x 6-foot high sections were the easiest to handle. I move my kennels every so often. This type kennel is great when your dogs get fleas or something, then you can pick the fence up and move it.

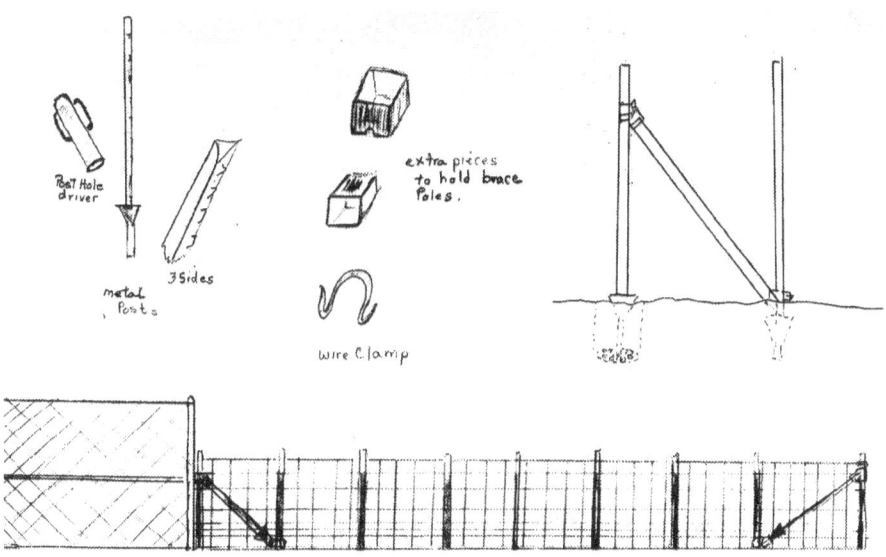

At the places where I didn't have enough sections, I put up metal posts that I purchased from the feed and seed store. I used a post hole driver (which is a metal bar with handles on each side and a large hole in the center). I put this over the fence post and slammed it down on top of the post. The post then drives itself into the soil. I had to do this in the month of March after a few days of rain had subsided so the ground would not be like cement. I used a garden hose stretched along the ground for a straight line. I placed my posts about four of my footsteps apart. That is about five or six feet, no further! I didn't cement them in as I would be taking them out one by one when I replace them with those heavy chain link sections. Next I purchased a kennel wire from the grange and borrowed a fence stretcher. I rolled the fence on the ground beside the posts then stood the fence up against the posts and hooked them over the top of the posts until I could get the fence pretty tight. I secured one end of the fence to the last post.

Don't forget to brace the last post with a horizontal post that leans up the top of the corner post and goes down to the bottom of the post nearest it. I used special brackets to hold this support post in place. Your hardware personnel at your local home store will help you to put that together. I had to do the same thing with the other end. All corner posts need bracing so the posts wouldn't loosen out. Next, I attached the fence puller to a bar that I sewed in and out through the kennel wire. I set the wire stretcher up on the corner post and then pulled the fence tight by cranking the handle. Make sure the fence doesn't get caught up on anything. When the wire was nice and tight, I clamped all the kennel wire to the posts with a strand of wire I had in the barn. You will need some wire cutters and you will need to twist the single wire tightly around the post and the kennel wire. Make sure the wire is well on the ground, even if it is a little too much on the ground. The kennel wire I used was only four-foot tall. I never encourage my dogs to jump a fence. Shepalutes are not too crazy about jumping fences, but if you have one that does jump, you might consider a heavy hog wire 6-foot tall and use wooden posts set in concrete.

The Dog House

Build the house with sleeping quarters three by five feet and three feet high at the highest point. Incorporate a front porch one and half to two feet deep and the five-foot width of the house. If the house is correctly situated, the porch roof offers shade from the sun and the porch itself is a place for your dog to get out of the rain or snow.

First make the skeleton framework out of two-by threes. Build the two side sections first and allow six inches of extra height on the uprights for floor elevation. Incorporate the porch size in the over-all length of the sidepieces and remember the back slope over the sleeping portion will accommodate the hinged roof.

Next, build the floor frame and cover it with five-eighths inch outdoor plywood. Cover the sides with the same material you use for the floor. If you allow your two by three inch framing to show on the outside of the house, you will have a smooth inner surface to attach your floor platform to. Keep the floor the six inches above ground level provided by your side uprights and brace the floor by nailing six-inch pieces of two-by-threes under the floor and to the inside bottom of the side uprights.

Frame in the door section between the porch and the sleeping quarters. The door should be four to six inches from the floor (to hold in the bedding) and twelve to fifteen inches wide with a height of fifteen or twenty inches. Animals like small caves to go into so don't get the door too wide. They should be able to squeeze in as an adult. My dogs all prefer the smallest holes.

Nail your plywood, or tongue and groove siding, over this framework, of course leaving the opening for the door, and nail the same wood across the back and the porch roof, thus closing the house in all around except for the roof section over the sleeping quarters. Build this section separately, with an overlay of four inches on the two sides and the back. Attach an underneath flange of wood on both sides and rear, in from the edges, so that the flanges will fit snugly along the three outside edges of the house proper to keep out drafts and cold. Hinge this roof section to the back edge of the porch roof and cover the entire roof part with shingles or heavy tarpaper, with a separate ten-inch flap stripped along and covering the hinged edge. Paint the house blue or blue-gray, as flies don't like that color.

In cold weather, tack a piece of rug over the top edge of the doorway to fall across the opening. An inside partial wall can be provided at one side of the door if you find the door is too large letting the cold in.

Make sure you build this thing in the same place you want to permanently place it, as it is big. I suggest you put this house at the door end of the run so the dogs will do their business at the other end. That way neither of you has to tromp through the waste. Try to set the house with its side to the North and its back to the west. The smaller and lower you can build your house, the warmer it will be in the winter.

Bedding can consist of marsh grass, oat, rye, wheat straw, pine, or cedar shavings. Cedar is said to discourage fleas and lice, plus it smells good. Shake a liberal supply of flea powder in the bedding once a week during flea season.

Flies

During the summer months, flies may bite the edges of your dog's ears. This can cause the formation of scabs as well as drive the dog crazy. A good liquid insecticide should be rubbed on the dog's ears. Skin disease salve, which contains sulphur and a small amount of Vaseline, may help.

The very best fly catching device I have ever used was The Big Stinky. If you get one make sure you pick up enough of that stinky stuff that goes inside. Every fly around will get caught in this trap without chemicals!

Travel

When traveling in hot weather never leave your dog in the car alone. Carry his water pail and food dish with you and take care of his needs at all the rest stops. I purchased my vehicle for the dogs and I leave all their stuff in it. I love to show off my dogs at all the rest stops. If you are ever traveling Interstate 5 and pull into a rest stop you may see me and my dogs!

You may want to begin his change of diet to plan kibbles a few days before the trip. Bring your own water in a rinsed out milk container.

If you must ship your dog, make sure the crate is large. I purchased the largest sized vari kennel. Put your name, address, and phone number on the crate in permanent ink. Do not feed your dog at least four hours before the flight and exercise him just before you say your good-bys. I put straw in the crate with chips of cedar at the bottom. The airlines ship animals all the time, so they will take care of him. Make sure the folks on the other end know the flight arrival time. Before your dog goes on this trip, make sure he has all his necessary tags and shots. You might want to make a copy of that and tape it to the crate. Do not put a lot of stuff in his crate with him. One item to chew on or something of yours might help him.

Chapter 10

Grooming

Mother Nature grows our Shepalutes coats to fit the weather and your Alsatian Shepalute is a hardy breed of dog. Exposure to the elements and regular brushing are a foolproof combination. If an Alsatian Shepalute is kept constantly inside and scrupulously clean, it won't develop much coat, muscle tone, or strong feet. Am Alsatian does need to come inside at regular intervals if he is to be your faithful companion dog. Socializing a dog means to give the dog a lot of different experiences. Slippery floors, carpets and sofas are necessary for him to learn about if he is to be that inside dog.

If your Alsatian Shepalute is to be an outside dog, farm dog, herd protection dog or outside protection dog, he will do just fine as long as you are around. If you intend not to spend much time with your outside or backyard Alsatian Shepalute I would suggest you consider two dogs, as one would be lonely. If you have two Shepalutes and do not allow them inside, they will be closer in character to the wild pack dogs. I would definitely consider neutering both of them. Spaying the female and castrating the male would calm down any behavioral problems that may arise.

Shepalutes do need brushing and combing and a bath once or twice a year, or when they get dirty. This is the time for your pup to lay on your lap or lay down and be still. This is the time to socialize and bond. Put your hands all over his body. He must trust in you, to give of himself completely. If you never bath or brush him, his coat will naturally shed around March or April when the weather changes and the days get longer. This event of nature regulates his coat. If you do not brush him, you will find him scrapping up against trees or bushes or fences to help pull the dead hairs out. All or most of his under coat will fall out. Once an old lady told me my dogs had "the mange" they looked so bad. A once over with the brush and they looked like they had lost 50 pounds!

Bathing

I bath my puppies when they get dirty or when I decide to bring one in the house for a while. My older dogs get a bath when the weather starts to change from rain and snow to spring and sun. My dogs are kenneled in five acres and are outside most of their lives. I put my 100 pound dog in a tub that is set up to drain out to my flowers so I don't get the hair down the pipes.

First of all, make it easier on yourself and brush or comb out all of your dogs hair one square inch at a time. On a large dog such as this one, it may take you a week. I use a dog comb that has a wide space between each tooth. This makes it easy to get down in under all that under coat. I part the hair in one-inch horizontal lines and I start with the back legs. I comb up into the rump and down the sides of the rib cage. Then I go on to the neck section, under the belly and then the throat. That is how the coat sheds. The shedding starts in the rear of the dog so the dead hairs come out easy. The neck section may not have many dead hairs if you began your grooming process earlier than the dogs coat has completely shed out. By the time you get to the throat section (two weeks from now) it will be easier.

Now you are ready to give your dog a bath. Properly secure your dog in the tub with a nylon leash and collar. If it is a warm day, tie him up in the corner of a yard or in the corner of his kennel run. You may need to screw in a couple of hooks. If you are bathing your dog in a tub, don't allow his head to extend out over the tub. It gets the floor (and you) all wet. Use a rubber mat so he will not slip or scare himself.

You may need to build a ramp if you have placed a tub in a garage or outside for the purpose of grooming your dog.

After the dog is well secured (at both ends if necessary) wet your dog from the shoulder blades back to his tail. A hand held shower nozzle will help immensely. If you don't have one, try a large pitcher. Soak the dog thoroughly, but don't wet his head.

I use an apple scented dog shampoo that has no chemicals. I dilute that shampoo with water and put it in a catsup container for easy pouring. I use it lightly, not heavily! Work the shampoo through the coat to the skin. Get every inch. Wait two or three minutes then rinse well to get all the soap out. Remember; don't put water on his head until the very end of the bath. Dry the dog's body as well as you can.

Now for the head. Some times I wash a dog's head with just a wash rag. (All dogs will shake when water gets on their heads). Never scare your dog with water in the nose. Lift his head up and put the water on his crown so it runs down the neck. He will now shake his whole body and you will be thankful you dried him as well as possible before you started to wash his head. Ring the water out of the coat and towel dry before he gets out. Walk him down the ramp and keep your body beside the tub so he doesn't jump over the edge of the tub.

Drying —- Retie the dog to a secure object to blow dry him or lay him down beside you. Again start with the back legs as you did with your combing or brushing. After he is pretty dry re-brush him and fluff him out. He is then crated in the mudroom (6x8) for about two or three days so I can brush him and love on him with his clean coat while we both watch television.

Cleaning the ears —- Put your finger in a towel and wipe the ear out as you dry the dog. If the ears are really dirty use a cotton swab. You may need to use an ear-cleaning product or use a small amount of antiseptic cream or you can put some alcohol on your cotton ball and wipe the ear that way. Do not put anything down in the ear. Do not go into the ear farther than you can see. Dry the ear by using cornstarch on a clean cotton swab. Do not allow the powder to drop into the ear. Just dust the ear really light to dry the ear or let it dry naturally.

If you suspect any problems let your veterinarian check it out.

The eyes —- I usually wipe the eyes clean with the face cloth when I bath my dogs. You may wish to use an eye cleaner for dogs. If your dog's eyes are red or gooey contact your veterinarian and your breeder to let them know. If you get soap in his eyes use diluted eye drops to rinse the eye out and/or rinse the eyes with water as soon as possible.

Manicure session —- I do clip my puppy's nails. At least four times before they are six months old. After that, I do not clip them unless they become too long. Maybe twice in a Shepalutes live time. My dogs are farm dogs and they all live outside. I also do not let my dogs jump on anyone. My dogs are to sit to be petted. I do allow some of my dogs to jump upon my out-stretched arm with both of their front legs to get petted so I do not have to bend over in my old age. The dogs only do this to the command "hup."

Use regular toenail clippers for small pups but make sure you don't clip the vein that grows on the underneath side of the toenail. If the nail does bleed use cornstarch or flour to stop the bleeding.

Conditioning the coat —- Good coat condition of any dog first starts in the genes. The next place good skin and coat comes from is nutrition. Lastly is brushing or grooming the dog's coat. There are many coat conditioner products in the stores; however, dogs don't care for aerosol sprays. If your dog's coat is dry he probably isn't getting enough oil and fat in his diet.

Dry cleaning your Dog's Coat —— if your dog doesn't really get too dirty, you may consider dry cleaning your Alsatian. Brush as before the bath. Get a trash sack for the dead hair you remove from his coat and wet rag for real dirty areas and the face. After you comb through your dogs coat use baby powder and cornstarch mixed together and shake it on his coat. Brush this in. Brush in both directions. Wipe his feet with a warm wash cloth. You may add some olive oil to the warm water and some perfume or herbs to make his smell better. Be careful. You do not want to use too much olive oil or mineral oil or you will have to give him a real bath to get the oil out!

You may wish to blow dry the powder out of his coat if you get too much in it. One last tip if you use a dry cleaner for dogs from a pet store it may have chemicals in it.

Doggie Odors and Colognes —- Your Alsatian Shepalute does not genetically carry a smelly gene. Seriously there are dogs that carry what I call the smelly gene. That dog just carries and odor about him and one can never wash it out. Some dogs have an oilier coat than others do. Alsatian Shepalutes do not carry either gene. They are a clean odorless dog. They just have a thick winter coat and they shed!

Every dog has anal sac glands on either side of his anus. In the Alsatian Shepalutes, these glands are emptied automatically when your dog does his business. Each dog's glands smell differently to another dog's and this marks his territory as well as urinal markings. Scant markings are stronger and the particular odor lasts longer to let other animals know whose turf they are in. If your dog's glands do become full and clogged, it would be uncomfortable. Some dogs scoot their bottoms on the grass or floor. It would be unusual for your dog to have any problems in this area, but if you are concerned about this, talk with your veterinarian and breeder.

The use of colognes or deodorants is a matter of choice. Read the labels as some may contain irritants. I prefer herbs or natural powders.

Grooming equipment

The comb —- This is the most important piece of grooming equipment that you will need for your Alsatian. Purchasing the correct comb for the job depends on the length of your dog's coat. Some Shepalute's coats are rather long and thick so a large comb with greater distance between the teeth is required. The distance between the teeth makes it easier to get into thick, long coats. You will find this comb does twice the job of any brush. Brushing usually only conditions the top hairs of your dog's coat and doesn't really get out that undercoat. An all-natural boar bristle palm brush will condition and stimulate your dog's natural coat oils. A natural bristle brush will also not put electricity into the coat.

Toenail clippers —- Use human nail clippers on pups. The pliers-type nail clipper is the strongest and best choice for the large dog. I only have to clip a nail here and there after my dog is about one year old. Usually the dew claws on his front legs. He shouldn't have any dew claws on his rear legs.

Thinning shears —- If I have a long-coated Alsatian Shepalute, I may need to thin the coat down on his rump or tail.

Small scissors —- Again the long-coated Alsatian may need scissors to clip the hair between the pads of his feet or the feathering. He may get tar in between the pads or a mat or two behind his ears.

Professional groomer's

If you take your pup to the Groomer, make sure you walk your dog before you go so he doesn't relief himself in their shop. Do not water or feed your dog before going to the Groomer.

Bring your own shampoo and leave these instructions.

1. Do not use any flea products or chemicals on my dog.
2. Use my shampoo please.
3. Do not clean his ears, I have already done that.
4. Do not clip the pads of his feet with a razor or clipper.
5. Do not use a clipper on my dog.
6. Do not clip his whiskers.

Something to think about —- Hundreds of dogs go through that grooming shop. Thousands of fleas, germs and other dog's hair are floating around.

Chapter 11

Healthcare

Warning! I am not a veterinarian! I give my advice per the constitution and in line with my experiences in life as I have seen it. I suggest you examine and study for yourself as I have done. I religiously believe in homeopathic and all natural remedies before man-made.

When you first get your puppy make sure you choose a veterinarian that you can rely on. I choose a veterinarian by finding out how long he has been in practice. If he/she came from a farm, all the better. Too many folks that say they love animals want to be a veterinarian when they grow up and have never even had a cat or a dog of their own!

If I have just purchased a pup I want to make sure that he is healthy. I purchase a new germ-free crate that I can haul around while the pup is going back and forth to the veterinarian's office. I make sure this pup gets all his proper shots and that the pup never touches any part of the veterinarian's office, except the examination table. I keep my dog in the van until it is our turn and I either carry him in or bring him in while crated.

I never take a young pup to a park or even in the back or front yard until he has got some shots in him and is protected.

After all his shots I still do not take my dog to any doggy park or any place where other infected animals have been. If I go to the park it is for a reason, training and socialization. If my dog belongs to a club and we have our trainings out in a field we go there to do our job and then get back in the crate in the van or back of the truck. Not much fun, am I? That's why I live on a farm. (Hehe)

Just try and be safe, both you and your dog.

First Aid Kit

I do have a plastic toolbox I converted into an animal first aid kit. Here is what I have in it:

A couple of rectal thermometers
Vaseline
Antiseptic liquids, sprays and creams
Burn cream medicine.
Bandages: an ace, a couple of rolls of sterile bandages, about 50 square bandages four inch by four inch.
Three rolls of tape for bandaging.
Cotton balls
Cotton swabs
Scissors
Two or three human eye-drop bottles that have been diluted with 75 % sterile water
Ear mite medicine
Ear powder antiseptic
Boric acid powder
Worm medicine
Shampoos and Herbal flea shampoos plus coat conditioners.
Brush and three different size combs.
Large human toenail clippers and two large dog nail clippers.
A vomiting liquid, to induce vomiting.
Cod liver oil and anti-stress vitamin B supplement
Tweezers
Hypodermic needle

I also have a microscope and slides, a stethoscope and various other possessions that I have accumulated over the years that may or may not be something you wish to get into.

Emergency first aid

Note again that I am not a veterinarian and do not profess to be. The following remarks are given as a result of my experiences in the field of dog breeding and emergency first aid that I have preformed on my own animals.

Normal rectal temperature for a canine is 101.5 °F.

Your dog will pretty much take care of himself, but if it makes you feel better to help him if he is hurt, then I shall suggest that you treat him as you would yourself. Wash out the dirt and apply an antiseptic.

Accidents

Poisoning —- If you fear that your dog has swallowed poison, get him to the veterinarian as soon as possible and try to locate the source of the poisoning. Take the bottle or container with you if you can or write the information down quickly. The container may say what to do so look it over.

Car Accidents — I shouldn't be writing this as I feel a car should never hit your dog! Train him to not go into the street, ever! Keep a leash on your dog at all times when he is not in the security of his own home. Read my book "Training your Alsatian Shepalute". If by some chance a car does hit your dog, he will be out of it, so to speak. He will be in shock and you may not be able to mentally connect with him, so speak his name and talk to him before you touch him as in this state he may turn on anyone who touches him. If this is the case, get someone to get you a blanket and keep talking to your dog. Reassure him until he let's you pet his neck and or hug him. Do not move him at this time unless he is harming himself by trying to get up. Wrap a blanket tightly around him and lift him into a waiting car. Lift a large dog with two arms. (If you are right handed, your left hand goes under his head and holds his chest. Your right hand goes under his belly). You are keeping the dogs body closely up against your chest. His feet are perpendicular with the ground at all times. Hopefully you will have help, as he is a large dog.

If he can walk, calm him with your voice, hold his collar (even if he is well trained) and slowly guide him to the car. Hopefully you have a car that was purchased for your pet and is roomy.

Pick your dog up with two arms cradling his underbelly and keep his feet pointing towards the earth. Keep his back straight and parallel to the ground. (This is extremely important for the animals since of security.) Place him on his feet and help him go down onto a flat surface. Get your drivers license and your keys and call the veterinarian to tell him you are on your way. Hopefully you can get your neighbor to drive your car while you sit next to your dog and keep him from moving.

If there is visible external hemorrhaging, if you're dog is bleeding, apply pressure. If you were able to get a dishrag or bed sheet or your emergency animal toolbox then you can apply that cloth tightly around the bleeding area (if at all possible) if not, just apply pressure. If an artery is pouring blood out of a wound you may need to apply pressure above the wound to stop the bleeding. Release the pressure at some intervals for a very short period, and then reapply pressure.

Blood coming from an artery is bright red in color and will spurt out in unison with the heartbeat. Blood coming from a vein is a darker red and it comes out in a continuous flow.

Burns and scalds ——- Call the veterinarian and tell him you are on your way. Keep the dog calm and apply a cool cloth to the burned area if he will allow it. You may be able to clean the burn by removing any foreign matter such as bits of lint, hair, grass or dirt. You may be able to spray cool clean water from a water spray bottle on the burn to cool the skin. Try to prevent exposure to air with a cool cloth. Do not use oil or butter! A semi-loose bandage may help, but not so loose as to move back and forth and irritate the burn further. Get to the vet. If the burn or scrape is minor, clip the hair away from the affected area and apply a paste of that burn antiseptic you have in your animal first aid box. Then apply a cotton sterile bandage and put the animal in an enclosed corner or vari kennel, watch him, as he will try to remove the bandage. Tell him not to do that or put an Elizabethan collar on him. You can make one out of cardboard or plastic. For your own legal protection take pictures and record the event. Record the heeling process.

Snakebite ——- Bleed the wound immediately. I squeeze the area and force it to bleed as I reassure the animal with soft tones and watch that he doesn't bite me. Wash the wound and apply tight gauze above the wound if you can. Get to the vet.

Bee Stings —— If you can see the stinger, remove it. Put one teaspoon of soda in a small cup and add a small amount of water to make a paste you can apply to the sting wound. Keep the pup still by holding him in your lap and petting him softly. If the pup wiggles, tell him no and continue petting slowly to keep him calm in your lap. Do not rub or agitate the area. He will be fine in about two minutes.

Infectious Diseases

In this chapter I hope to give you a little bit of information that will enable you to make more informed decisions about your dog.

Bacteria, viruses, protozoa and fungi (which invade the body of a susceptible host and cause an illness) cause infectious diseases.

Infectious diseases —- They are often transmitted from one animal to another by contact with infected urine, feces and other bodily secretions, or inhalation of germ-laden droplet's in the air. A few are transmitted via the genital tract when dogs mate. Others are acquired by contact with spores in the soil, which get into the body through a break in the skin.

Bacteria—- They are single-celled germs, while the virus (the tiniest germ known and even more basic than a cell) is simply a package of molecules. Although germs exist virtually everywhere, only a few cause infection. Fewer still are contagious. Many infectious agents are able to survive for long periods outside of the host animal.

Antibodies and Immunity —- An animal that is immune to a specific germ has chemical substances in his body's system called antibodies. Antibodies attack and destroy that germ before it can cause an illness.

Natural immunity exists, which is species related. A dog does not catch a disease, which is specific for a horse, and vice versa.

If an animal is susceptible to an infectious disease and is exposed he will become ill and his body will begin to make antibodies within himself to fight against that particular germ. When he recovers, these antibodies will afford protection against re-infection. They continue to do so for a variable length of time. He has now acquired what is called active immunity.

The reason wild animals never need vaccinations is because in God's breeding program, animals that cannot fight off infectious disease die. Those that fought off diseases and lived acquired active immunity against those diseases. Giving our domesticated animals infectious diseases that have been watered down, so to speak, creates antibodies to form within that animal's body. These antibodies attack and kill that watered down germ. That is what happens when we vaccinate our animals.

Now if that animal ever came into contact with that same live germ while the animal is playing in the park, that animal's body would again form more antibodies to help out those antibodies already in his system. Together, they would overcome the germ. During this time when antibodies are forming and fighting off germs, your animal will need plenty of rest, clean water and proper nutrition.

There is another type of immunity. It is acquired from one animal to another from their mother's milk. If the dam were never vaccinated against a disease, her pups would receive no protection against it.

Passive antibodies can stop a vaccine from stimulating that puppy's active immunity because the mother's immunity has overcome the vaccine passed on to the pups from the breast milk. This is one reason why vaccinations do not always take in very young puppies.

Distemper Vaccine —- The Distemper vaccine that your veterinarian gives your dog may be a killed germ or a modified live germ. This first distemper shot should be given shortly after weaning and before a puppy is placed in a new home where he will be exposed to other dogs. A high percent of puppies do not get a satisfactory take from a distemper shot due to circulating maternal antibodies. The measles virus is similar to the distemper virus and it is not so affected by the maternal antibodies that travel to the pups from the mother's milk. The measles shot is able to stimulate antibodies, which will also protect against the distemper virus. Booster shots are required. I do not give my future-breeding pups the measles shot. I only give this shot to pups that will be sold as a pet. The measles shot has interfered in some of my female's reproductive abilities.

Hepatitis —- Adenovirus preparations (cav-1, cav-2). This vaccine protects against hepatitis and the adenovirus, which is a virus implicated in what we call the kennel cough disease. The hepatitis protection is included in the DHL shot that you give a 12-week-old pup. The letter H stands for Hepatitis. Booster shots are required every 2 1/2 years, though vets like to give your dogs booster shots once a year. (Go ahead ask him how long a DHLPP shot lasts). If he lies to you, I'd find another vet. There is also another virus that is present in kennel cough called Bordetella.

Leptospirosis —- Leptospira bacteria will protect your pup against the two types of bacteria that cause leptospirosis. The fist shot should be given at three to four months of age. Leptospira is incorporated into the DHL shot. The letter "L" stands for leptospirosis. Booster shots are required. Again every 2 1/2 years.

Rabies —- There are two general types of rabies vaccines. One is a modified live virus preparation, which means that it is a modified live virus. The other is an inactivated virus. Rabies vaccines must be given in the muscle. Care must be taken to be sure the product is made specifically for dogs.

When I was young, rabies shots were given to all dogs when the dog reached six months old. A study was taken that found out that most puppies died by the age of six months, so by changing the law to require all dogs to get rabies shots at the age of four months the state could increases their profits two fold. I called the health and safety department in the state of Oregon to find out when the last dog ever found to have rabies was. She told me that there had not been a single case in a period of ten years. I am thinking that if it were longer than that she would never have told me. I also think that ten years may be as far back as her records show. (For a reason). It is suppose to be public knowledge. Anyone can check up on this type of information. Go ahead call your state health and safety department and let me know what they tell you.

The rabies vaccine has changed. Now the rabies shot will protect your pup up to a period of three to four years.

Rabies shots provide the county and state with all their information on pet dogs within each jurisdiction. These statistics are public access. You may call or write your county health department to find out when where and how the last rabid dog was found in your district. When your veterinarian gives your pup a rabies shot several forms are filled out. One goes to the county health and safety department, one goes to the county animal control and the veterinarian keeps a copy. This way everyone keeps a record of how many dogs each person owns. How old the dogs are, where they live and if they have a county or city license.

County and city licenses vary in cost. If your dogs are not neutered the price is very high, around 80.00 per year. If you neuter your pet it can cost you from 100.00 dollars and up to 300.00 in veterinarian charges, but because you neutered your pet, your county or city license will only cost you about 20.00 per year.

In most areas throughout America, you may have up to three or four dogs without having to obtain a kennel license. Some counties have only a few kennel licenses and most all are taken.

Is it bad for a four month-old pup to receive rabies shot so young? In my experience, the pups that received rabies shots at four months have not taken the shot as easily as a mature pup.

Parainfluenza-Kennel Cough —- Bordetella bronchiseptica vaccine is available and is of aid in the control of another agent implicated in the kennel cough complex. Two initial vaccinations are given and require booster shots.

149

Parvovirus —- Both inactivated and modified live virus vaccine's result in effective levels of parvovirus antibody. Initial vaccination consists of two doses given three to four weeks apart. Annual boosters are required to give your pup the needed protection. For maximum protection in high-risk areas, vaccinate at two-week intervals until the puppy is sixteen weeks of age.

Who needs maximum protection? If you are taking your dog to a lot of shows or puppy classes or if you visit a whole bunch of puppies during your day then you will know that your pup will require extra protection.

My Vaccination Schedule

5-8 weeks Canine distemper-measles-cpi

14-16 wks DHLPP (distemper, hepatitis, lepto, parainfluenza, parvo)

12 months DHLPP and rabies

Every 3 yrs. Rabies

Brucellosis
Leptospirosis
Tetanus (lockjaw)

Note: actually the DHLPP shots last for a little over 2 years, but veterinarians thought it was better to get boosters into the public's dogs every year. They knew we would forget and this would keep the dogs relatively safer.

Viral Diseases

The following is a list of Canine Virus Diseases:
Distemper
Hard-pad
Herpes virus of puppies
Infectious canine hepatitis
Rabies
Infectious Tracheobronchitis (kennel cough)
Canine parvovirus (cpv)

Fungus diseases
Norcadiosis, actinomycosis, Cryptococcus
Histoplasmosis
Blastomycosis
Coccidioidomycosis

Protozoan Diseases
A Protozoan is a microscopic single celled organism that is mostly aquatic and which includes many parasites.

Toxoplasmosis
Trichomoniasis
Giardiases
Piroplasmosis
Coccidiosis —- This is an extremely common protozoan disease found usually in young dogs. Puppies can acquire the infection from contaminated premises or from their mother if she is a carrier. If your pup came off a farm or has been in and around any streams, cow manure or chicken feces then your pup might have this disease.

Your veterinarian will be able to tell you if your pup has contacted coccidiosis by a stool sample that you provide him at your well puppy checkup.

Here is a little bit of background on this disease:
Five to seven days after the ingestion of oocysts, infective cysts appear in the intestines and feces. The entire cycle is complete in a week.

The first signs of infection can be a mild diarrhea, which progress until the feces become mucus-like and tinged with blood. There is loss of appetite, weakness, dehydration and anemia. Often a cough, runny nose and a discharge from the eyes will accompany this. It does kind of resemble distemper.

Coccidia can be found in the stools of puppies without causing problems until some stress factor such as an outbreak of roundworms or some kind of stressful situation occurs. This reduces their resistance. Dogs that recover can become carriers. They remain in good health, but can suffer relapses when afflicted with some other disease, such as distemper.

Finding adult oocysts in a microscopic slide of fresh stool can identify carriers and dogs with active infection.

Treatment: Take your pup to the veterinarian and/or get a stool sample checked. I have often just taken a stool sample in to the vets in a plastic container with my last name and dog's identification marked in felt on the container. My veterinarian charges me $8.00 to check the stool. If my pup is found to have cocci then I take the pup in for a full check over and treatment plan. Your veterinarian may suggest you stop the diarrhea with a heomycin kaopectate anti diarrhea preparation. A severely dehydrated or anemic dog may need to be hospitalized for fluid replacement and blood so make sure you get that pup checked out at the veterinarian's office!

Supportive treatment is important since, in most cases, the acute phase of the illness lasts a few days, perhaps ten days, and is followed by recovery in uncomplicated cases.

Sulfonamides and antibiotics have been used to treat coccidiosis. Response is slow once the signs of disease are apparent. Known carriers should be isolated and treated. At the same time their quarters and runs should be washed down daily with Lysol and boiling water to destroy oocysts, otherwise they will re-infection themselves.

Nervous System

Fits (seizures, convulsions) —- A seizure is an uncontrolled burst of activity which begins with champing and chewing, foaming at the mouth, collapse, jerking of the legs and perhaps loss of urine and stool. Then there is a brief loss of awareness followed by a gradual return to normal. A blow to the head may bring on seizures. Seizures may have been inherited.

Hypoglycemia —- Can cause seizures as well as coma. I also believe this is a hereditary trait.

"Worm fits" —- May appear to be a seizure to you. Your pup may go into a fit during heavy infestations with intestinal worms. The cause of seizures is unknown but it could be due to low blood sugar or serum calcium.

Common poisonings—- Which induce seizures are strychnine, lead, insecticides, and rat poisons.

Brain tumor —- If your dog has an inherited brain tumor he may show signs of confusion, aimless wandering or some other change in the behavior of the dog before he goes into a fit. Other signs are an unsteady gait, loss of coordination, staggering and enlargement of a pupil. These signs may appear more as the tumor grows.

Bee stings —- May cause your dog to go into a frenzied barking followed by fainting or collapse. If you didn't realize your dog was stung, you might think he was having a seizure.

The Distemper virus —- Begins with champing, tongue-chewing, foaming at the mouth, shaking of the head and blinking of the eyes, then a dazed look. The pup soon gets worse and dies.

Epilepsy —- Epilepsy is a recurrent seizure disorder, which originates in the brain. When it is due to a blow to the head, distemper, or bacterial infections of the brain, it is said to be acquired.

When it is due to genetic disfigurement of the brain or when the cause is unknown, then it is said to be congenital.

Congenital epilepsy —- I believe this to be a genetically transferred disease as it has appeared more in the St. Bernard's, German Shepherd Dogs, Poodles and Beagles than in any other breeds.

Epileptic attacks are recurrent and similar. An epileptic seizure has three phases to it. The first is recognized by the onset of sudden apprehension and restlessness. He may cower and try to hide in the bushes or underground. If he is in the house he may go into a corner. There may be bizarre behavior, such as sniffing in the corner or snapping the air or just staring into space. Next, the dog will start foaming at the mouth and his eyes may flicker. He may start a chewing with his mouth. During the next phase the dog collapses or lies down and his body goes stiff. His pupils may dilate and he doesn't appear to know where he is or what is going on around him. The dog appears to be coming back into his surroundings. He may loose control of his bladder or bowels.

After the seizure seems to disappear the dog remains confused, scared or uncertain. If over stimulated by a loud noise or rough handling, a second seizure can occur. The first two phases pass quickly (in about three minutes). The post-seizure state can persist for several hours. This might give the impression that the seizure was of a long duration. However, a true epileptic seizure is over in less than five minutes. Seizures are genetic. Do not breed this dog.

Treatment: There were several dogs that have come into my grooming shops that had epileptic seizers. I have also owned a couple of dogs that had seizers and I have found the following to be true.

If your dog starts to have a seizure, stand aside until he quiets down or cover him with a blanket to make him feel safe. If a blanket upsets him don't cover him of course. (Don't put your fingers in his mouth or try to wedge something between his teeth.) I would move the dog into a very quiet and calm environment then call your vet. If petting him and talking calms him down then do that. I have found that even my calming effect is too much for a dog that is having a seizure. The best thing I know of is to place this dog in a small area where he will feel safe. Keep everybody quite. No noise and no other pets around him. The seizure will be over in a few minutes. Your veterinarian may want to examine the dog to exclude other conditions that might be the cause of this seizure. Seizures lasting over five minutes are dangerous. Permanent brain damage can occur.

A number of drugs are used to control or prevent epileptic seizures. If you have a dog you believe has seizures please consult your veterinarian.

It is my belief that nutritional supplementation may help. I am positive that they have helped with my dogs. The dosages recommended here are for my own adult dogs that weigh in over 100 lbs. Again, please consult your veterinarian!

Essential

Vitamin B complex	100 mg.	3 times daily

Extremely important in the functioning of the nervous system

Extra b3	50 mg

Improves circulation and is helpful for many brain-related disorders

Vit. B6	100-600 mg	3x day

Needed for normal brain function

Folic acid	500 mg.	Daily

A brain food vital for the health of the nervous system. Anti stress Vit.

Magnesium	700 mg	Daily

Divided doses calms the nervous system

Very important

Calcium	1500 mg	Daily

Nerve impulse transmission

Zinc	50-60 mg	Daily

(Do not exceed 100 mg.) Protects the brain cells.

Helpful

Kelp 1000 mg
Alfalfa
Vit a 25000iu
Vit. C 2000-7000 mg
Vit. E 400 IU then increase to 1600 IU daily.

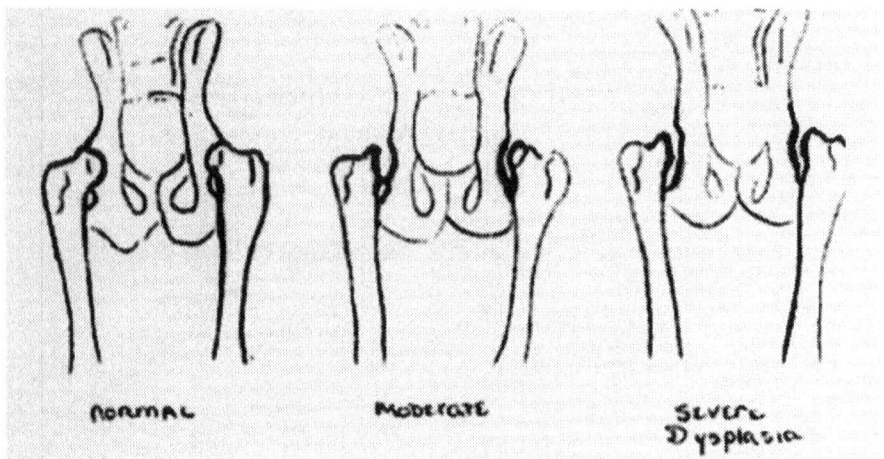

Hip Dysplasia

This is the most common cause of rear-end lameness in the domesticated canine. Large breeds tend to come up with hip dysplasia more than the smaller breeds. I have had dogs with varying stages of hip dysplasia. Some of those dogs no one would ever have known had this problem. The following might help the reader to understand this problem.

Sublazation —- means there is a partial dislocation or an incomplete dislocation of a joint. An example would be the partial dislocation of the femur and pelvis.

The femur or head of the thighbone should sit solidly in the socket. But if the hip socket is shallow the femoral head will be able to slip in and out. With this slipping in and out of the socket the head of the thighbone becomes worn or flattened and the problem intensifies.

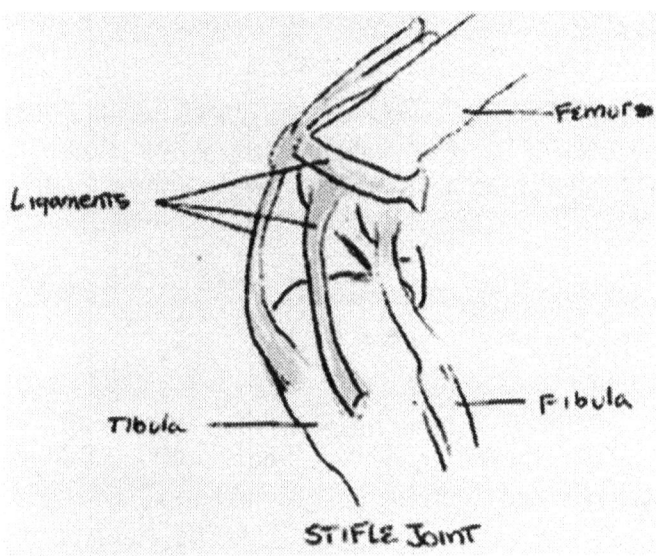

Osteochondritis —- This is an inflammation of both the bone and the cartilage around the joint.

Cowa Balga —- This is defined as a deformity of the neck of the femur bone and is the worst bone problem I can think of. I believe this to be an inherited deformity, worse than a shallow hip socket!

When the femoral head of the thigh bone begins to move from the shallow hip socket the gristle like capsule that surrounds this area thickens to hold the femur in place then calcium deposits begin filling the gap. That area becomes inflamed and arthritic conditions appear. Eventually it reaches a peak and the capsule breaks under stress and then the animal must be destroyed or undergoes surgery.

This condition usually becomes apparent between the ages of 3 to 12 months. I have never had a dog that had this severe problem.

By breeding for tight ligaments and a broad pelvis with a lot of muscle mass, I believed that this might help the Alsatian Shepalute stay away from inherited hip dysplasia. Of course, not breeding to animals who show signs of hip problems such as not wanting to raise their front legs up onto the water buckets at a young age, or not wanting to jump into the bed of a pick-up truck, all help me when determining which dogs to breed.

Aseptic Necrosis

This is where the head of the femur bone doesn't get enough blood supply and this leads to the gradual destruction of the hip joint.

The signs are severe lameness and refusal to bear weight on the leg. These same physical signs may end up to be something totally different than aseptic necrosis as only an x-ray will be able to diagnose this problem.

Elbow Dysplasia

This is a faulty union of one of the elbow bones called the anconeal process to the ulna bone. A build up of cartilage or bone may end up floating around the elbow joint. This will be painful and will cause abrasions. This also occurs during the growth period and can be detected in pups about six months of age. This inherited developmental problem primarily affects German Shepherd Dogs.

Panosteitis

I have found this inherited disease within the German Shepherd Dog lines while breeding my Shepalutes. Panosteitis was never seen in any of the other breeds that I have breed. My first encounter with this problem came in an eight-month-old pup I had sold. I had always tried to keep in contact with the new owners, especially in the early years as I monitored all the outcomes of my breeding. The owner of this particular pup had called me and we talked about our dogs. When she mentioned that her pup was limping on its front leg, I made a special trip out to her place to check the pup out. I checked the dog all over to try and find any sores or maybe even a thorn, anything to let us know why this dog was limping. The limping got so bad that the pup cried and would not put that front leg down. The veterinarian took some x-rays and told us that, in his opinion, this dog had panosteitis. That was when I did the research. There is no cure for this problem. The limping on the right front leg ceased and the limping now changed over to the left front leg. How weird is that? Then, that cleared up and two weeks later a limping began in the right rear leg, then the left rear! Of course I never bred that dog. I did breed others in direct lines out of those dogs' parents and of course all pups were constantly watched. No other cases came up in those other lines. Of all the pups that I have bred only five that I know of had this problem and all were connected genetically. I also have found that if my pup showed these signs at a young age the condition was the most sever, but if an older pup showed some limping it was not so bad and would just be a short period of time before he stopped his limping.

Here is a way to tell if your limping pup has this problem. Apply pressure over the bone shaft of the affected bone. The pressure alone will produce pain. X-rays will show a density in the bone. It is said that the growing spurt causes this, and that the pup's bone marrow hasn't caught up to the bones growth. I personally have seen this to be so when I have bred a thin boned dog to a larger boned dog of the German Shepherd Dog lines. I also found that I could help such a pup out by giving this pup food high in the B vitamins along with calcium and bone meal.

Pituitary Dwarfism

Canine pituitary dwarfism can happen in many breeds, but it is most predominantly in the German Shepherd Dogs and so I thought I would add this problem to this book.

For the first two to three months of life, the pup with this problem may appear normal and indistinguishable from its other normal littermates. After three months, you will see that this pup isn't growing like the rest of the litter. You may notice that this pup's coat is shorter than the others are and that the hair's it does have will have a tendency to fall out. The adult hairs don't come in, in contrast to the other pup's coats.

This pup may also have behavioral abnormalities such as fear biting and aggressiveness or just a weirdness about him that you can't put your finger on.

In one of the early breeding's, a really nice Alsatian Shepalute threw some runts whose coats were shorter and thicker and the pups turned out really nice. No such problems as this pituitary dwarfism have ever occurred in any of my breedings. By the way, I did not use that small dog in my breeding program, as my standards call for a large dog. It did occur to me though that I had seen some very nice looking small German Shepherd Dogs during my childhood years. I tried to find some research on them, but never did find out where they came from. I doubt that they had pituitary dwarfism but such a genetic characteristic is just what begins a new breed!

Skin Disorders

In my 40 years of working with dogs, I have seen many skin problems. I would like to tell you that all the skin problems I encounter have appalled me! I am a witness to the anger and confusion of the public who owned these dogs as to what the problem was and how to get rid of it. I have seen owners spend thousands of dollars trying so hard to replenish the coat and to stop the consistent itching and biting as well as to try to get rid of the smell that always seemed to come hand-in-hand with this type of problem. I have heard stories as to what the veterinarians thought these problems were and I have seen thousands of products being used on those dogs that claim to be the cure when nothing seemed to work. These problems attached themselves to no one particular breed, though the English Bulldogs and Sharpies seem to have it the worst. Cocker Spaniels get it bad, too! The ears get infected so badly that the veterinarian has to surgically close the ear. Of course, the dog then goes deaf in that closed ear and Lord, what a smell those cocker ears have. I owned a couple of cockers that came up with ear problems, a weeping yellowish fluid that continued to run out of the ear and into the hairs surrounding the ear. Like so many other cocker owners, I spent a lot of money trying to help these dogs. One day a product came in the mail with a fully refundable guarantee if I wasn't satisfied. It was the only thing that has ever worked. It cleared those cocker ears up in three days! I swear by it, but the veterinarians and cocker owners lectured me on putting a powder in my dog's ears!

What is the problem? In my studied opinion, if it isn't worms (especially tape), and it isn't the poor food (nutrition), then it is an inherited genetic deformation that is of course, not wanted. Simply that. Poor breeding!

In my breeding of the German Shepherd Dog, I had some litters that contracted skin problems. A persistent scratching with reddish spots and loss of hair. Sometimes a change of diet and epidermal tar shampoos would help and/or I would find the dog had tapeworms, which I believed to be the underlying problem. A great imported male shepherd out of Germany came to me with a persistent ear problem along with a skin problem that showed itself mostly during the summer. He would rub his backside up against anything that would reach his top line near the base of his tail. His hair would begin falling out and his skin would scab over from the biting he did. I kept a constant watch on his stools during this time and sure enough tape worms! We never could get rid of those worms in this particular dog. I went so far as to shave the hair along his spinal bone approximately eight to twelve inches long and directly above the onset of the tail. Then I would shampoo that spot with a tar and sulfur shampoo and leave it on for about five to ten minutes. Then I would rinse the coat clean. Four days later, I put an Iodine shampoo on and I did the same thing all over again. Of course, I would enrich his food and cook his brown rice. His skin would clear up in two weeks and his coat would come back in nice and thick. During this same time as his skin problem, his ear infections flared up, which turned out to be a yeast infection.

The removal of the tapeworm along with the application of the ear medication and the increased vitamins and minerals as well as the sea kelp and applications of tar and sulfur shampoos all helped in the healing process. His hair grew back in and the itching subsided until the tapeworms again appeared! We lived in the mountains in the snow, so fleas were not a great problem. By the way that cockamamie theory about fleas being the reason for tapeworms is for another book! I have never believed tapeworms came from fleas and to this day, I do not!

Anyways, I did breed this dog, as he was too nice not too. A very smart dog and so extremely devoted! My calculations were to mix the gene codes by out breeding. All his pups were closely watched and only one line (four generations later) ever had bad skin problems. Sporadically, other pups did get slight problems with their coats. Yes, it is inherited. In the last ten years, we have had an absolutely clean slate with these problems. None of the pups from those lines were used to continue into this breed, though. I made sure I flagged all his prodigies' folders and kept a special eye on them. By the way, the veterinarian told me the dog was allergic to grass. Another veterinarian said it was summer itch!

One of the third generation (F 3) puppies I sold began a skin problem as early as six months of age. A smelly reddish skin that itched the poor dog. He continued to loose his coat further and further down his neck then around his face. The owner of that dog returned the dog to me and I gave her a new pup from another litter. I had been monitoring this particular line of shepherd x malamute mixes and had noted the fact that some of this line had similar skin problems but none so dramatic as this pup. When this puppy contracted this skin problem I was not too astonished, though I did not think that this might happen as none of the dogs in the upline had any problem. This was proof enough to me that skin problems were inherited though, as I have stated previously. It seems to be very recessive and only when two special dogs are bred together would it show up. But that's for another book!

Within my breeding practice to develop this breed I must be adamant about the breedings! I must refuse to breed with any dogs that turn up with skin problems, ear problems, bone problems or any other problems.

In the first generation with skin problems in the genes, 2% of the puppies acquired skin problems not seen until six months or older. By that time, I had chosen a nice male to breed with for the next generation. This stud dog only had a slight skin problem but after three more generations, it came out in those puppies full bloom! That pup that I have described in the previous paragraph was one of the offspring. I had no other choice but to eliminate that particular line completely from my breeding program. I strongly believe that if all breeders would only use the best pups many existing problems within the purebred dog world would cease to exist.

In the beginning years of this breeding program, I had formed over forty different lines in order to breed back to them. Today I have only two lines from the original stock left! Today in my dogs I have no so-called summer itch, no eczema, no hip problems, no eye problems, and no ear problems. Now I'm not saying they may never pop back up. I stress that all new owners keep in touch with me so that I can continue to keep an eye on any problems that may arise. Only then can I adjust my breeding and document the offspring on the parent's records.

Day Blindness

This condition can be detected in puppies from seven weeks of age when the breeder will see these pups bumping into objects and not being sure of distances during the daylight hours outside. Curiously, during the night and in the house, they see fine. It is only outside in the sunlight that this particular pup seems to bump into objects.

Hemeralopia is associated with the enzymes which supply the chemicals that transfer the light that goes into the eye and into the nervous impulses of the electrical energy that is carried along the optic nerve.

It is believed that this recessive gene has been inherited down from sled dogs of the North. This condition has also been noted in poodles. I myself have never seen this, but because this condition seems to appear in breeds that are recognized as coming from the North, I felt I should include it. I have never had any eye problems in the Alsatian Shepalutes.

Chapter 12

Training your Alsatian Shepalute

In this book (and in this chapter) I will give a quick overview of the training necessary for your Shepalute puppy to be that dog you have always wanted! For further and more complete written instructions on how to train your Shepalute please read "Training your Alsatian Shepalute."

Most of the puppy training books I have come across relay the message that If I am not a good trainer/parent and If I don't socialize, bond, and prepare my puppy correctly then it is my fault that the pup is so hyper, digs up the backyard, or barks a lot. Nonsense! The problem is the breeding. A human being cannot stop an animal from being who he is. One can train or modify an animal's behavior and that can mold the animal or suppress the animal into an animal that a family can love and tolerate. Certainly a normal man or woman who knows nothing about behavioral training will throw up their hands and break down in tears when they have to bring that eight month old dog to the pound. It is not all your fault. You are only human and most dogs are not meant to be apartment, city dwelling, and humanized, well-controlled and behaved pets! I have seen it all too often and I have seen the family go through one heck of an ordeal. That is why I started breeding these dogs. I got tired of hearing "this dog was abused." Why? Because he is shy? Most all the time the dog was bred that way, because folks don't understand that breeding is an educated art. Shyness is dominant. But at last there is an answer, there is hope for dog lovers and you will soon understand what I am talking about, as this dog of yours is so easy to train. You don't have to be an animal behavioral specialist to raise an Alsatian Shepalute!

The puppy's first year is the time for the owner to show and teach the pup your rules. You shouldn't have to try and change the dogs herding, coursing, running, digging or barking characteristics. You should spend your time showing the pup how to fit into your household.

Puppies are not born knowing your rules. They are not born understanding the English language either.

In order to train an animal, the animal trainer must begin the training in the correct brain mode. First, the trainer must consider the animal to be trained. Since this is a book about the Alsatian Shepalute, my comments on training will be directed specifically towards this breed, the Alsatian Shepalute. All professional trainers know that different breeds are trained and handled differently. This is because they all work differently. They all have different sensitivities and characters, which make them all, respond differently to different types of training programs. It is difficult if not impossible to write a dog-training manual that would cover all dogs.

Praise and Punishment

Next, we must take a look at all the different training methods. All the different books published on how to out-think and trick the dog into doing what you want it to do. Forget all that gibberish. We are pretty smart human beings; let's talk about how to get your pup to behave as a companion dog around the house and around your children. Let's get him to behave on walks or when we take the kids to the park. It's not all so difficult. Why make it so hard? First, find out what your dog wants and let him have what he wants if he performs the habits, tricks, or behavior that you want him too! So, correct instantly and reward good behavior with the warmth and reassurance of physical contact along with a slice of hot dog! Hot dogs? Why not? Animals have always learned certain behavioral habits that humans wanted them to exhibit by rewarding them with food.

I know an awful lot of you are upset at me now for even suggesting that you use food rewards.

Please just trust me on this.

It will not make your dog into an animal that only does tricks for food! Nor will it spoil his appetite and make him a finicky eater. But it will get your dog to watch you, your hands and listen to the sounds you make. He will follow you everywhere you go anticipating, waiting with patience only an animal out for a chicken frank can have! I keep packages of sliced up chicken franks in my freezer for training. An animal will do anything for a food reward! Later they will do for you because they have learned, but until they begin to understand, why not make it easy? I personally guarantee that food reward training is the easiest, simplest and the most enjoyable form of training that I have applied for any animal that I have ever run across!

Remember this, no matter how inadequate your dog's performance is you must end all lessons with the puppy doing something right and a food reward until the lesson has been learned. And if you have a problem, stop the training. Go to the Alsatian training book and figure it out. If your mood gets in the way or your pride gets hurt because your pup didn't do something right, stop. Get over it and think about how you can make it work and how you can make it happen the way you want it to. You are the problem, not the pup. You are also the one with a human brain so put on your thinking cap and find a simple solution.

Keep your lessons short, about 30 seconds or one minute. You can gradually increase the lessons as the pup's attention span grows. I know you are in a hurry. I know you want it now. But have patience. Time works.

Most Shepalutes love to do things for you. They like your attention and the praise. Shepalutes will figure things out pretty fast if you show them what you want, and then right after that, give him that hot dog slice. Shepalutes are not easily distracted especially if there is a food reward.

Ok, we have the praise and reward down, but what about punishment? And do we have to? Even if you do not believe in violence or force, there will come a time in your life when some one or some dog challenges you. You are about to get hurt. It is a fact of life. Nature and this world does not care if you understand or don't understand. Nature does not care if you get hurt or killed. Nature and the world keep on turning and life goes on whether you cry or not. Nature itself is force and violence.

All animals (including humans) challenge and fight, beat up, or chase away opponents. Just keep in mind that any amount of force that emanates from you must be in direct context to the amount of bodily harm you believe will be applied to you personally. Are you going to get hurt? Is your life in danger? You have a right to protect yourself. But if you can get out of a situation that might cause your death you must do that first and above all. That's the law. No preemptive strikes unless your at war with Iraq.

In the dog world, dogs bite and fight and may even kill those that interfere with them and what they want. Your dog is an animal. Animals do not get together and make up rules or laws to govern themselves by. They don't get together first and say, "Ok now, no biting in the eye, or no grabbing you-know-what." I have seen the way animals carry on and that is the nature of things. Dogs bite with teeth that tear the opponent's skin from the bones. Dogs snap those teeth closed and a puppy goes scampering. Pups get bite marks for life. One of my cocker moms snapped at one of her pups and took out the pup's eye. Now you know she didn't say, "Oops your eyes look kind of bad there, better call a human right away." Nope. She didn't even care, because she couldn't. Animals don't care. Sorry. They don't have that capability.

Punishment—-There are many ways to punish and there are different words I could use to say the same thing. How about compulsion or negative stimuli? I'm not one for mincing words so I'll just stick with punishment. You can punish a dog by merely looking at him or by using a severe tone of voice or by actually bopping him on his behind or scaring him with a loud noise. Stopping a dog from doing a bad act can be a very hard thing to do. My chickens or goats for instance! Sometimes the life of one of my animals is on the line. Now I know from experience that physical punishment is not what stops my dog or pup from killing that animal. Here is the secret, it is fear. If I can put the fear of terror into my dog or pup for even looking at that baby chicken then he will ignore it. He will actually turn his head and look up to the sky! He will slowly get up and crouch away. He will go somewhere that the baby chick will not get close to him. I am telling you it is not the amount of force or hurt or pain that you apply to the animal's body. It is the noise you make. It is the confusion, the amount of action, and the newspaper swatting as hard and as fast as a whirlwind onto the closest object then again on the other side of the dog. Smashing a tree and the branches swish over head and a bunch of twigs go spinning and the world seems to be falling down around this pup. His master is yelling and stomping and crouching and my gosh, that newspaper!

Again, I am talking about the hard dog. The Alsatian Shepalute is not a touch sensitive dog, but he is sensitive in nature. (So to speak). He will not like all that noise and confusion. One must of course judge how much of this to use depending on the emergency and the totality of the problem. I rather think it is a very bad thing for one of my dogs to eat one of my chickens or goats. It cannot be tolerated. It is one of my rules. A big no, no. These dogs easily learn not to displease me. Let me clearly state again that this is not physical harm. None of my dogs go to the veterinarian's office with broken bones let alone a broken toenail!

One can also use the many different tools or hardware that are out there on the market for consumers to purchase such as leashes, chains, chokers, pinch collars, halty leashes, etc. All of them apply pressure to the animal's physical being. But how much force is enough is gauged by the response and the desired reaction of the animal being trained. If the animal continues the bad act, the amount of fear did not overcome the dog's pleasure in performing the presumed bad act.

A professional trainer has experience in many different techniques and levels of punishment and rewards that will get the response he is looking for. Whatever is called for, a professional handler knows that the emphasis should always be on the timeliness of the punishment, not strictly on the severity involved. The severity means nothing if the trainer or owner does not catch the pup in the act at the time. And a professional trainer sets the event up for the pup to do the act so that it can be dealt with. This is the major difference between a human that has never owned a dog and a professional handler. I strongly advise new owners to join any and all dog classes in their area. Experience different methods and techniques.

A trainer knows the value of praise and rewards. And of course not enough can be said about praise. Keep in mind that the praise level must be higher than the punishment level at all times.

99% of the problems in training a dog are the owner or trainers fault for not following through with his words. I call that lying to your dog. Lying is a habit a human being gets into without realizing it. Little words or white lies (cotton tails).

Here are some Examples:

A grandmother tells her grandchild of two years old that her face is dirty and that she is going to wash her face off. This is a lie. The young child is scared now because she thinks her grandmother will wipe her face off and she will no longer have a face.

The dog is not allowed on the couch. You leave the pup alone unsupervised while you go to the store and the pup gets on the couch. You have lied to him. You told him he couldn't get on the couch, he could, and he did.

You tell the dog to sit and the phone rings. You go to answer the phone. The dog didn't sit. To the dog, you are a liar. Sit doesn't really mean sit, especially if the phone rings.

You tell the dog to lie down after you have trained the dog what the word "down" means and you also use your body language for the word "down." The dog looks at you until you get mad and your voice changes and your body language tells the dog you finally mean it. (I've seen those dogs actually count the amount of times a student tells the dog to down!) Hehe.

Those are just some examples so you could understand how these little lies confuse or distort your true communications. That's what lie's do, even little funny sayings that us grown-ups don't pay attention to.

What does all that mean and how do you stop yourself from lying? First you must think more of yourself. You must know that words are powerful. You must believe that a word is who you are. Once you finally realize that you are your words and that words are not to be taken lightly then you will think about the words that come out from your being. Each word counts and each word matters. Every time you utter a sound that sound travels at the speed of sound out into the universe for eternity. Do not say what you do not mean!

It's that simple. If you cannot enforce the word you utter or make sure it is followed through, then don't say it.

More on this very important subject as we go through each of the different commands.

Socialization and Informal Training

Socialization is the education of your puppy in the everyday events of life. Your pup has never seen the world before and it is up to you to show your pup what the world will throw at him. Noise, cars, vacuum cleaners, can openers or just walking down the end of the street and back will bring many different sights, sounds and smells to your new puppy. He is so young he does not know all these things, so in the beginning take your dog absolutely everywhere you go. Carry him in his crate or cage in the car and remove him whenever you stop to window-shop. (Make sure he is up to date on his shots).

It is extremely important that this pup does not ever do anything that is deemed bad by you. You hold the rules and he must learn them. You need to let him know what the rules are. And you must not change your rules. (Lie to him). Your rules must be written in your blood. (So to speak)

Laundromats are good places to bring your dog while the clothes whirl around in the washer or dryer, then you and he can walk around the block and amble back and forth in the parking lot of the crowded shopping center. Be sure that all his inoculations are completed before you start early training anywhere but on your own property.

Try to locate nearby handling classes. If you have joined a kennel club, it probably sponsors a class or two and they will give you the details. If you cannot find a class, check with your breeder, the newspaper or your veterinarian.

I realize that a lot of people feel that a ten to fourteen week old puppy is too young for class, but I disagree. Work your puppy just a few minutes at a time. Bring his crate along and allow him to sit in it and watch the other dogs while he gets used to the confusion and noise.

This is how I look at all this training stuff: Say for instance World War III breaks out and your life and your pup's life are on the line. You see a life-threatening situation and you yell to your dog to stay there, your dog stops and stays there. Then you say "down" and the dog downs and a car crashes into the path your dog would have been on if he were not trained.

Now maybe that scenario will never happen to you and your dog in your lifetime, but what if it did? I know my dog would live. Would yours?

To experience life totally, one must live life as a warrior. Always being prepared for anything. Be aware of your surroundings. Slow down and listen to the world speak to you. Be as one with your dog. My dog is an extension of myself. He is my arm, my hands, my eyes and my ears and I feel safe when he is near.

House training

Shepalute puppies have a strong natural instinct to avoid soiling their own area. If you are consistent and pay attention to your pup and his times, you will have no problems. Shepalutes do not want to potty in the house. If your pup does potty in the house, he has not learned how to get out. Where is the outside? Which is the door? Have you placed him in a huge house so that he gets lost and confused? You are responsible for teaching the pup how to get outside. He was not born knowing the rules or being able to speak to you nor does he possess the ability to make a map of your home and to be able to read it. He also cannot open the door by himself.

If your pup is eight weeks old, or older, your pup has learned to potty in fresh straw or grass if he came from my kennels. He will be upset if you do not let him know where to go. Where is the grass? He needs to be shown. If you know he needs to relieve himself right away, do not let

171

him walk across the floor a long distance to the out side. Pick him up and carry him to the outside. Watch his brain work as he figures it out the next time he feels the urge. He will want to go back to the same place, the same door, unless you have him in a mansion of many rooms. Then boy, will you confuse him!

Q- How do you stop a pup from peeing on the rug?
A- Don't let him on the rug!

Remember that your pup will need to go every time it wakes up. (What goes in must come out). Five a.m. is when all Shepalutes wake up. When the first rays of light appear over the Eastern horizon and you are cuddled in your warm bed having a great dream. And what is the first thing you do when you wake up? Go potty?

Wild animal babies explore the area around their dens first. As they get older they will begin to venture out further and further until they get to know the different smells and the direction back to the den. Let your pup get used to the room the door is in first. Let him get acquainted with everything in that room. That is the room all his stuff should be in. His food and water dishes must not be next to the door to the outside. If they are, you will not be able to tell whether your pup wants a drink or wants to go potty. Animals go as far as they can in the opposite direction away from their food and water to relieve themselves. As they get older, the will go still further. My adult dogs go to the hog wire perimeter line. Usually in about a 10 x 20 foot corner. They will make a complete route around the property and when the hog wire stops them from going any further that's were they will go.

Open the door every time your pup heads in that direction. Always, even if you just sat down! Your pup will understand real soon and will wait for you to sit down before he heads for the door! Tip: Crate him when not supervised. No water after six p.m. Exercise will make a pup go potty. Knowing this will help you in your house training. Getting excited will stimulate his body organs, be prepared.

Here is a secret that I have learned. Perhaps it will help you in understanding your domesticated animal. "The larger the area you harbor your pet in, the less he can expand into later and the wilder the animal will become. The smaller the area the calmer and more bonded or tame an animal will become."

One must also remember that the area your pet lives in has to fit the animal's needs. The area you keep an animal in must give him space to relieve himself, eat and rest, get up and turn around. And of course it depends on how old that animal is and what your intentions for that animal are.

For instance, say I want my Alsatian Shepalute to be a well-mannered protective pet dog and to bite when threatened or on command. This means I intend to really put some work into this pup and that I must make sure I know the laws! This pup has a purpose. This pup has a job to do and that is how I look at this pup. There is my mind set.

I must begin this dog's training while still a pup; say 6-8 weeks old. Everything I do will be towards that ultimate goal. In this instance my pup will live in a crate while he is young and if I got him at six weeks old, perhaps a playpen to start out. The reason being is that I want total and complete bonding. This pup will need to look to me for everything he needs. (For further instructions on training a protection dog, please email me at:

Shepalutes@aol.com

What if I want a pup to just be a farm dog? Then I would gate off the mudroom, which has at least one door to the outside. That will be his room. This mudroom has tile floors for easy cleaning and disinfecting. I would put newspapers down in the corner closest to the door. I would make the enclosed area about 4x6 feet if the pup was six weeks old, so he will not feel overwhelmed. I would spread the cage out as he grew. I would also make sure that anything and everything within that caged area was safe for him. He would also have a toy box full of things to keep him busy. And I would rotate his toys from the toy box to the cage.

Belonging to the Alsatian Shepalute Club will provide you with written material pertaining to this breed of companion dog. Click on the web site (Members.tripod.com/rarek9) to review and update yourself on the breed and or e-mail the club directly will give you further information. This is the most intelligent way to become established, oriented, exposed, instructed, and protected as you learn about the breed with other Shepalute owners and you get hands on look at your pup's relatives.

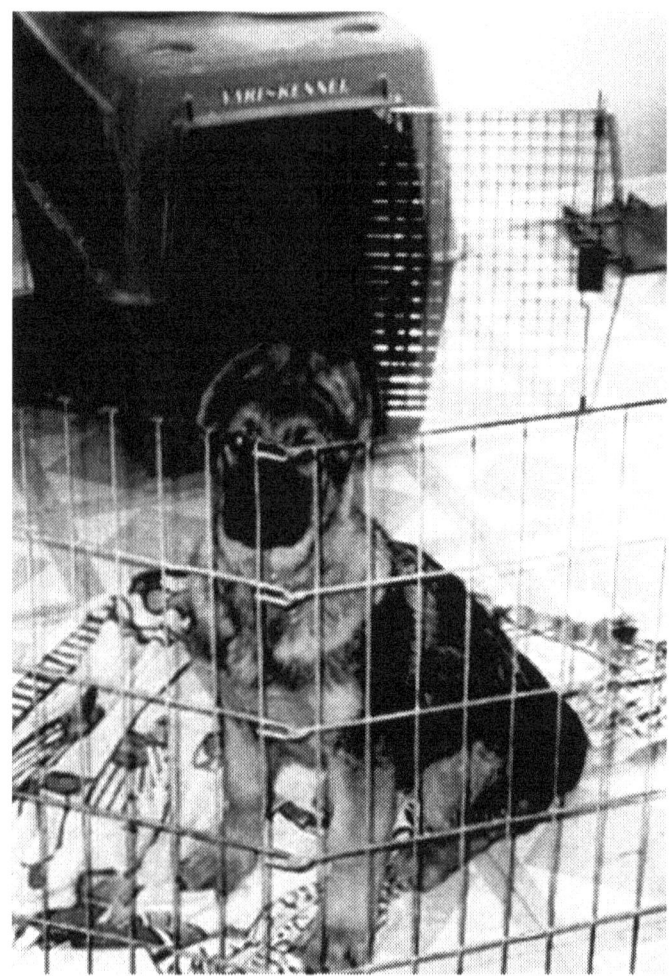

Vegas (short hair) 2003

Crate training

I am assuming that you chose a pup to be a member of your family as 90% of those who have bought my Shepalutes have gotten them for this purpose. If you wish to train your pup for attack or protection please read *"The Alsatian Shepalute Protection Dog Training Manual."* You can order that manual by e-mailing me directly at (Shepalutes@aol.com.) If you wish to train your pup for show, field trails, herding or herd protection again, get the manual for your particular lifestyle and training purpose.

Since this pup goes with me everywhere that means he is not unsupervised. My eyes and ears are upon him at all times, even when I stop to talk with a friend. This is almost impossible, you say? Well, yes. So when I cannot or do not wish this grave responsibility, the pup goes inside his cage or vari kennel. Now the vari kennel goes with me and stays near my feet so I can tap the cage to let the pup know I am here. Another thing I might do is to tie the leash to a piece of furniture in the room where I am working. This is an important bonding period.

All pups need to investigate. They need to know where the food and water dish is, where the door is, and where their crate is. You may want to give him only small amounts of water during the night for easier potty training until he settles in. Remember what goes in must come out.

Shepalutes like caves or sheltered areas to hide their food or toys and to stay warm or safe.

Pop your pup in and out of his crate with some small tidbit for him to chew on while he is inside his crate. I throw small pieces of hot dogs inside the crate with a command "over" or "kennel" depending on what this dog will be assisting me in when he is older.

I have found that the easiest way to train a Shepalute as a housedog is to use an exercise pen while he is young. This prevents him from getting into trouble, forming bad habits and allows him more room than a crate. I spread newspaper all over the pen. This pup can then relate the feeling of the newspaper on the pads of his feet with that of relieving himself and that will give you a head start on paper training while in the mud room. Make sure that when he finally does come out into your living area that no one in your household forgets to pick up the weekly newspaper from off the floor!

Exercise and Feeding Program

While any animal is in training his food and exercise program should be figured out to best benefit the training that your pup will under go. If the training is physically strenuous he will need the proper nutrition. Up his proteins and the "B"vitamins for stress.

Make sure that after the training sessions you calm him back down. A horse gets walked and brushed out. A dog gets walked and petted to calm him back down. A marathon runner walks and bends over then walks more and breathes deeply.

I do all my training before a dog gets fed. Keeping your dog a bit hungry will ensure a more positive, energetic, "want to do"attitude. Of course I am not talking about starving a dog. Just watch the dog's reaction to the food reward and the type of foods you use to stimulate his interest in the lessons.

Shepalutes do not need a lot of exercise. They do not need a lot of room. They are heavy boned dogs and would rather pull a heavy load than to run five miles.

"Go potty" on Command

Exercise and training go hand-in-hand if your dog is a city dog. Take your pup out at the same times every day. Go the same route and let the pup potty at the same place. Bring a plastic poop bag just in case. All my dogs potty on command and its easy to train your dog to do the same. Feed and water him well, then bring him to the area where you want him to go and let him sniff naturally. As he lifts his leg or she squats say "go potty" in a soft voice. When he/she finishes praise him. While walking your pup, do not allow your pup to potty wherever it wants. There is a place for that and a command. This will be a real pain in your side when your dog gets older and out weighs you or pulls your arm out of its socket. Let your pup potty before the walk, at a designated turn around, and/or at the end of the walk. That should be plenty when you go on a twenty-minute walk. (Twenty-minute's when a pup is older, only a few minutes to start with).

While walking my stud dog that weighs over 100 pounds, I need to make sure he knows we are on a walking exercise. So first of all, before we even go for the walk my stud dog is allowed to run around the farm to relieve himself. Then we have a special leash and collar for pleasure walking. That means he doesn't pick fights, he doesn't sniff continuously or lift his leg every five seconds! He is to do that ahead of time, before we step foot off our property. After we leave that invisible property line the chance for marking territory or messing around is over. No more lifting the leg until we get back on our property. That is our rule and there cannot be any leeway.

Plan the length of your walk according to the age of the pup and its training level. My four-month-old has never been on a leash, so I put on his flat collar and a 12-ft. leash. He understands the come and sit commands and as he runs in the opposite direction I call him to me. I watch his stress level as we head out of the property and down the road. It is unfamiliar territory so I walk about 80 feet all the while calling him and then praising the pup when he is close to me. 80 feet out and 80 feet back. End of lesson. It took me twenty minutes to get the pup and myself ready and it took us eight minutes to walk our first planned heeling exercise. No go potty was used, as it was inappropriate at this time. (Reminds me of fixing dinner, two hours to prepare dinner and five minutes to eat it!)

Come and Sit

1. Always tell your dog to sit. Don't tell a dog to sit and then say, awe, I guess you don't have to right now, that is lying to your pup.

2. The Golden rule for the sit command is: Never pet your dog unless he is sitting. Don't let anybody else pet him unless they tell him to sit either. The moment your dog gets up from a sit, walk away. Turn your back. Don't pay him any more attention. When you are sitting on the coach and your pup comes over and looks in your eyes and wags his tail, maybe even nudges you- do you automatically pet him and smile? Well don't!

3. Never feed your dog unless he sits first.

To teach a pup to come and sit takes all of two minutes.
This is how fast my puppy learns to sit:
I go to the refrigerator and pull out the baggy that has chopped up hot dogs and cheese in it about the size of a dime.
I rattle the bag and kneel down as I call for my puppy in a high pitched voice. If he doesn't come I go to him and kneel down about two arm lengths away. I hold out the hot dog slice and let him sniff it, but I clinch my fist as he tries to get it and I bring my hand in towards me as the pup follows. I continue saying, "Vegas, come!" "Good boy" and I open my hand as he eats the treat. Before he goes away or jumps on me or tears my hand up or eats my fingers I stand up and get another treat. When he looks up at my hand (he knows there is a treat there) his bottom will automatically go down. If the pup is really hungry, I continue giving him treats (cheese this time) as I speak his name until he gets full of the cheese

177

and the excitement is not so bad. He has calmed down and realizes the cheese will continue to come if he just sits there. Voila!

I move across the kitchen. He waits a bit then follows me. I then tell the dog to sit and bring out the hot dog piece. I hold the hot dog slice just above his head and slightly back towards his shoulders so that he sits. (Since a dog is parallel to the ground he must sit to look up). As quick as a mouse I give him the hot-dog slice. Take your time and get another and repeat the same step. If the pup tries to get the hot dog, close your fist. He must sit first.

Don't put your hand so close next time, but don't let him jump up either. It's a bit tricky but pay attention and work with your timing. Stop before your dog is full and tires of the game. Do this once every hour for a period of one or two minutes. Two more times and your pup will never forget it. Now when the refrigerator opens or a bag is opening your pup will come running! Hehe!

While he is following you around the house and he figures out you have hot dog tidbits, hold one out for him to sniff then push your hand back over the middle of his head. As his head tries to follow the hot dog his butt will sit so he can get a better view. Give him the hot dog immediately, before he jumps for it. If your pup is jumping for it, your hand is too far away.

Come

This is the most important command in the world! This command may save your life, your dog's life or the life of your child. This command is not to be taken lightly.

1. Never lie to your pup. Never tell your pup to come and then not mean it. This means you should never use the word "come" unless you mean it. If you do use the word come and your pup does not come, you must go get him. Don't keep asking him to come with no result. If you do, he will realize that you don't really mean what you say. You should get a leash and choke collar first so you can be in control and your pup will not be able to get away. Go get your shoes on or your coat and hat, it doesn't matter if it takes you a while, just don't get mad. If you are mad and out of control, wait. Count to 10. If your pup does get away, shame on you! Now you have a pup that will run away. He wins! He will now have to live his

entire life with a rope or leash attached to his collar. Next time when you call him and he doesn't come you will be able to win.

2. Never punish your pup for coming. If you are going to punish him, you must go to him. Always. Get the leash and choke chain first! If you accidentally tell him to come and he does, you must praise him no matter what he has done to upset you. Come is always safe. No matter what. It's that important.

3. Never tell your pup "come here" and then give him a shot, bath, or medicine.

Look at the come command this way: Your body is the safest spot in the world for your pup. When your pup is near your body he is protected and safe, unless your body goes to him, loudly and stiffly and the voice that emanates from you resembles an earthquake, dark clouds or a whirl wind. That is the way punishment, for an act that was done by your pup that is not allowed, is dealt with. It is an "act" put on by you. It is the act of anger that the pup reads and understands. I don't even have to be angry, but the pup perceives me to be. Your pup will never forget what brought that upon him. Note that all those bad things your pup does must be dealt with the same way. If your pup gets away with what you consider bad, he will learn that he can do the bad act when you are not around. You are the human and you must put on your thinking cap. You must make it so that the pup does not do any wrong unless you are setting the pup up.

When your pup comes when you call him, go the extra mile. Get up and go to the fridge and find something to reward him with every single time. Believe me food works. He will love you for it.

Down

While you are teaching your dog to come and to sit, pay attention to your arms and hands and your basic body language. Your dog has begun to read you. That's how you speak to him. Some humans don't know this, now you do. (Hehe) Your pup should be watching your hands by now. This is very good! The down command is your closed fist with a hot dog in it going down to the floor in front of his nose. You may push your pup down on the floor and give him that treat immediately as his elbows touch the floor or, you may do this down command an easier way! A way that he won't fight against!

Wait till your dog is lying down and place a treat between his two front feet. "Good dog." If he gets up, walk away. If he stays down, give another treat, one after another as quick as you can before he gets up. This is the reason your tidbits must be small. Remember if he gets up, walk away, and pay him no attention.

I prefer this command. I love this command! While in the house I always treat a dog or pup that is lying down. The dogs do not understand it at first and will get up and come to me for another one. "Oh my, she's giving out hot dog slices for no reason!" "How strange is that?"

My dogs pick it up rather quickly though and they lie down and look at me to see that I have acknowledged them lying down. Just walk over to a lying down, calm dog and pat him on the head and tell him that you approve of this type of behavior. If he gets up while you are approaching him, walk away and say nothing.

Here's a secret, Always acknowledge that they are lying down! Don't forget that please!

Stay

This is another very easy command! First start with that vari kennel. I start this stay command at the age of eight weeks old or when I first acquire my pup. The lesson is repetitive vocal sounds emanating in a firm tone from you out into the atmosphere. When I put the pup in the vari kennel, I give the command "stay." When I walk away I say the command again. Then I come back with a treat. Then I go away and then come back. I never leave the pup for more than 2 minutes in the beginning. As the pup gets use to me coming and going he hears the word over and over again.

I never go back to the vari kennel with a reward if the pup is whining and carrying on. I go back to such a dog with the sound of hard feet on the floor, my body stiff, and my voice upset and low. I yell at the bad dog and may hit the top of the vari kennel. Then I say "Quiet." If he doesn't get quiet, then I loose! Now I will have to not pay him any attention until he becomes quiet.

I repeat the stay again and hide myself around a corner. I count to five or ten and before the pup cries I return to that "win-win" outcome. He gets a reward for being good. (You as the trainer must get back before the pup whines). I don't emphasis the stay command if we have a problem just being in the vari kennel. If your pup is crying and whining in the vari kennel then your pup has a problem being in the crate. You must give that problem your full attention before you go on to the next command. When training any animal you must train for one thing at a time. Get that one thing done first. Work on one problem or command at a time. Don't confuse the pup.

As you walk away from the cage that you put your pup in, tell your dog to "stay." When you come back praise him, give him a reward through the cage door. Continue doing this until your dog is about four months old. Guess what? He now understands the word "stay", sort of. Of course every time you walked away from the playpen, as he was older two and three months, you also used the word "stay." I also hope that you have watched your body language. I don't care how you do it, just do it the same every time. My stay is my raised up arm that is perpendicular to the floor and my index finger pointed straight to them. My eyes are then fixed upon them looking them directly in the eye and my voice is deep and stern.

My pup has just graduated to the exercise pen. It is a folding fence about 3 feet high and 10 feet long. I can complete a circle with this fence or I can use it in the mudroom up against two walls forming a small caged in area.

When I put the pup in this cage, I say stay and walk away. When I return the pup gets a treat. I never treat a crying, whining dog. I note how long I can be away without him crying and add ten seconds more the next time. Reward and praise as always.

Whenever I leave my pup I use the word stay or stay there. I want the pup to know that I will be returning. That is what that command means to my dogs.

I never use this command if I will not return. That would be lying to him.

I never use this command if the dog will not stay. I always make sure that the command will be followed or I do not say it.

If your older dog is tied up next to a tree or in the back of your truck or tied up anywhere else that he must stay at, then you may use the word "stay."

Voila! Simple, huh? What's that you say? Of course he must stay! Why yes, that's the way to train him. He wins, I win. This is not a formal stay as in training for an obedience trial. If you want that kind of a stay then you must train in that type of environment and with military proficiency.

Next step is to put your dog on a leash attached to a post and tell the dog stay, hide yourself then go back and praise the dog. He will learn that stay means you are going to return. Use the word "wait" if you will be calling the dog to you or ordering the dog to do something.

The following phrase will help you to remember which words to use in training.

Stay until you die.

Wait until I call you.

Of course your dog will never stay until he dies. It is just a saying that will remind you that your dog is on a stay and you must return to him to take him out of the permanent stay that you put him in. I once placed a dog in a doggy bed and used the word "stay" and in about thirty minutes I past his bed and saw him there! I went over and praised him and used the "ok" to tell him he could leave if he wanted. I really don't think the dog was obeying the command, or was he?

Wait until I call you

This command is used on the recall and is explained in greater detail in my book Training your Alsatian Shepalute. This command also has its own body language so the pup will understand more thoroughly and doesn't get confused. The only problem with body language and a command is that you cannot do either in an A.K.C ring, but then again I don't figure you will be doing that, or will you?

Heel

Heeling my dogs outside the obedience game is not the same as inside the ring. In this book we are not training a dog for the ring. So my rules on heeling my dogs are simple:

Do not pull on the leash. Walk beside me, slightly behind me or a little in front of me, but do not cross my path as I might trip. Do not cross another human beings path as they might trip. Do not disrespect other person's properties. Do not walk on someone else's grass. Do not walk in the street ever! Sit when I stop unless you feel you are threatened and never sniff anybody especially in the crotch. Ignore all people and animals unless I introduce them to you. Never growl or bite unless you feel threatened and if you ever do feel threatened I will remove us both to someplace where you will feel safe. I will respect your feelings and you will respect mine. I will protect you from harm and keep you out of negatively charged conflicts and you will protect me when I can no longer protect myself.

Remember that these are my rules. You must also have rules. You must write them down so you know them. As it is, you probably don't have any and have never thought about it. Now I ask you, how can your dog know the rules of his life with you if you, his master, do not know your own rules? You first must know the rules of your life. When you know them, you will convey them to your dog, believe me he will know.

Once the puppy feels at home, he must learn to walk on a leash if he is to safely negotiate the world beyond the front door. Let your pup drag his leash around the home for a while until he gets used to having something attached to him. Only with supervision. Don't let him get hung up on things and pay attention so you could help him if he needs you.

1. Don't let the dog pull the leash while you are holding it.
2. Let the pup run around with the leash and collar on while supervised
3. Quickly jerk the leash when it becomes taunt so that it loosens up.
4. Praise the dog with words and pets when it walks on a loose leash on your favorite side
5. Go for a lot of walks together.

Now we can begin our heeling exercises———-

Since my dogs are farm dogs the perimeters are clearly marked with electric fencing, as I do not want my dogs to ever go off the property. My dogs are not anyone else's problem. Nobody else chose them for a responsibility.

My dogs do not leave the invisible line that crosses the driveway into the dirt road in front of our country farmhouse. I must make that clear to all my dogs. They are never to go off this property. I am adamant about that. I cannot protect them if they go off the property and many things can happen. Someone can instill fear and anger in my dogs by kicking them or throwing things at them. Yelling at them, or even shooting them. If I allow my dogs to roam the neighborhood then I have placed strangers in control of my dog and his training program. Those strangers are now able to influence my dog's character.

The dog off your property may break your rules and once rules are broken, bad habits become carved in stone. Your dog may now cross the street, chase pray, destroy property, bark, run, tease, and even bite. All that is irresponsible and against the laws of most states. Such actions created by you ultimately cause more problems in life. Who needs more problems?

So I begin my heeling exercise with a heeling collar and the heeling leash twelve feet long. I call my four-month-old pup to me, reward him, and then snap on the leash. As he runs away I call him back to me before he gets to the end of the leash. If he gets to the end he will punish himself with a great force around his neck. He will be in shock, restrained, and captured! I will call him again to get his attention off the pain and onto my voice. I pull the leash to give it slack with a quick jerk and release. Then when restriction subsides, I bend down and call him to me in a high pitched happy voice. He remembers that I am safety and love and hugs and comes running to me for protection. I stay there with him for about a minute petting and reassuring him that I am his safety. I get up and begin to walk as I talk to him and ask him to walk with me.

Now the important thing here is that you don't get that entire leash wrapped around yourself or your pup. You must not let that leash pull his neck while he is safe by you as you are walking. There must be no restriction on his physical being. You must be his safety. A place where there is only love and every time he is with you he must never get spanked or feel threatened. Later in his trainings you may use a form of punishment that will communicate you are displeased but he will still know that you are his safety. So before you get up and start walking. Unwrap all that leash and roll it up loosely in your right hand. If your dog bolts the leash should be able to unravel easily until he again comes to the end of the line. He will learn quickly that twelve feet is a boundary line and he will begin to slow down before he gets there. You will then call him back to you. Jerk and release the line to get his attention. Then praise him when he gets there for about a minute to reassure him. As you heel your pup for the first week, only give him one rule. Don't pull the leash.

As he learns this and his stress level drops, start bringing in the twelve-foot boundary. You may also start guiding him to your left side by praising him every time he is on your left side and ignoring him or giving him some bad experiences at any other position besides the left side. Do not praise him when he is on the wrong side!

You are closing the circle of his freedom and guiding him to your left side. Make sure nothing bad ever happens to him when he is on your left side. Do not step on his feet or run into him. You must stop moving or go faster. If you do stop, bend down and guide him to a sit on your right side then you can pet and praise him. Never pet or praise him anywhere else but your left side.

Make your heeling exercises short at first. My fist heeling exercise was about forty feet then back home and a sit. Then an "ok" to finish off the exercise. I then let him run off to the kennels to pout.

As your dog begins to heel you will need to start a more in depth body language communication so that your dog and you can communicate back and forth as a team. We are going to pretend your dog is deaf so that you can more easily understand the importance of body language. Clap your hands and wiggle your fingers as you run backwards and bring your hands in toward your stomach as your pup reaches you. This is the beginning of the command to come or the recall command in plain body language. No verbal command is necessary!

Once your dog is comfortable heeling on your left side begin heeling by stepping off with the foot he can see. (Your left foot.) Walk in a straight line at all times and walk in a pace that keeps his attention. (Usually a fast pace). Your pace must be consistent. That means don't slow down and don't speed up. If it helps to put a hot dog in your left hand or just the smell of one on your shoe do it.

To Stop Your Heel— stop your heel by letting your dog know you are going to stop! You are a team and heeling is not a trick to be known only by you so you can prove your dog stupid. You are a team and he must win. So how do you do that? Remember I said that your pace must be consistent? Ok, three steps before you stop you will tell your dog you are going to stop by slowing down your pace! A change of pace will alert your dog to "something is wrong, or something is about to happen." You get his attention right away! Now smash your left foot down firmly on the ground as you stop. It might startle him at first, but soon he will know you are at a dead stop and what comes after that? He will sit. Since a dog can hear the change in footsteps your last step need only grab your dog's attention not scare him. As he watches for this "stop" to come after the change in pace, you will have a team effort in your heeling exercise that will always and forever be a silent communication between best friends!

Right and Left Turns—- Dogs read everything you do, so watch your body movement at all time.

Place an x on your left heeling shoe to remind you that your pup watches this foot closely. When you turn right or left show the pup you are going to do something by slowing down your heeling pace 3 steps before your execution. If you will stop or turn right or turn left, make the x shoe be the one that tells him what you are doing. Use that foot first when executing your turn.

Point Mugu naval base basic obedience class.

Dog classes

The fact that the dog in a dog class is more excited or difficult to handle is no mystery. This is the reason dog classes are invaluable when you and your dog are ready for distractions. All these distractions teach your dog to mind no matter what, to pay attention to you. Distractions also teach the handler how to make the dog pay attention by making us use our brains and of course that hot dog slice!

Some dog class instructors may be pretty new to the dog training game. Some instructors will refuse to allow you to use treats. Each instructor is different in how they train their dogs so pay attention to your trainer to see if this trainer is the one for you. If the trainer is a bit tough with a brass and hyper dog, don't assume the trainer doesn't know what he or she is doing. I have seen hyper Schutzhund trained dogs that have needed harsh punishment by a professional trainer and the dog has bounced right back. That is what I call a hard dog. The trainer knew his dog and the dog knew the trainer. With this kind of dog or even a Rottwieler who weighs 120 pounds you better be able to knock that dog down and/or be respected by the dog, if not, you have no business owning that particular breed or type of dog. In America, the choice is yours and so is the liability when your dog does whatever he wants to.

If the dog class trainer uses that same hard technique on all the dogs then you do not need this trainer's direction.

In this book we are talking about the Alsatian Shepalute. If you have any other breed of dog and are reading this please know that this is not intended for your breed of dog. All breeds train slightly different as the personality and sensitivity of each breed is different.

Your Shepalute puppy should go into a puppy training class or a basic obedience class as soon as possible. The reason I say to join a class is for your own experience and training. You will learn much more than the dog! Hehe. But seriously, if you join a class your dog will get use to being around other dogs and will not fear strangers. Its good training.

If you allow your Shepalute in the back of your pick-up truck, please put him on a restraint that is not long enough so that your dog could put his feet on the side of the truck. He is not allowed to put his feet up on the sides of the truck.

1. It will scratch your truck.

2. Your dog might loose his balance and go over the edge and hang himself.

So keep the leash short. Just long enough to allow the dog to turn around, sit, and lay down. When you come to a permanent stop then you may untie him. I personally don't. I also do not allow my dog to jump over the side of the truck bed when we get to our destination either. He must wait until I lower the tailgate and wait until I use the command "ok" which means we are finished. My dogs only jump into the truck when the tailgate is down and I use the word "hup."

Barking

Barking is an inherited trait!

It may be a dog's life, but it is a man's world. Your dog, as well as my dog, must live in this world under the rules that govern the human society.

Owning a Shepalute will help, as Shepalutes do not bark much. They do howl at the moon in the middle of the night, though!

New owners always call me very worried that they have never heard their dog bark and ask if this is normal. Do not worry! Alsatian Shepalutes do bark but please do not ever encourage a dog to bark unless you are in a competitive sport that requires it. At that time your dog would learn to bark on command. Know that a barking dog is a scared dog or just a neighborhood nuisance. A barking dog can be compared to the little boy who cried wolf.

Barking is hereditary as well as the sound's a dog makes. This sound that comes up from his belly and throat is what identifies him to all other animals within hearing distance. My dogs can tell one another's bark. I also know the sound of my Alsatians. Their barks are so similar that a Shepalute cousin down the road sometimes confuses me. They sound so much alike.

The Alsatian Shepalute's bark is deep and low. A high-pitched bark or whine is not acceptable in our breeding program.

Digging

Excessive digging is an inherited trait!

Why do dog's dig?

1. Because they learn to dig. Another dog taught them or you did. You laughed and petted him when he dug, or you allowed it or set it up to be so.

2. The weather conditions are so hot that a dog will instinctively dig in the shade to reach cooler dirt. It is a survival instinct. Don't allow you dog to require cooling himself down. When it is, hot spray the yard as well as the trees so that they act like a cooler when the air blows over the leaves. Wet down the cement. Bring your dog into the air-conditioned home.

3. Your dog is a female and it is about four weeks after her heat cycle. This is a normal survival instinct. Most female canines dig dens to whelp their pups.

4. Your dog is bored and there is nothing left to do.

5. Digging is also an inherited trait. If one breeds dogs that do a lot of digging then you will get pups that do a lot of digging. The ground dogs were bred especially to dig and to enter dens or holes in search of prey.

Rafael and Suzzanna 1980

189

Jumping on you

Jumping on you is a learned behavior!

It is also a genetic trait in that it is part of hyper ness.

If you allow your dog to jump on you it becomes a habit that the dog gets into to get your attention and it has worked. This is a dangerously bad habit!

Jumping on people and other members of the pack is normal in all canines. This is how they begin to play which elevates into a dominance thing. If you are not in control then your pup is. When a canine is in control it can be extremely dangerous for all concerned. Besides that, it is extremely rude, impolite and might cause the owners of that jumping dog a lawsuit!

Jumping on you starts when the pup is young. Teach the dog to sit and never pet or give attention to a dog that misbehaves. It is not cute.

If your dog is already a jumper than you have already set in this behavior. Now we must try and correct it. A guru once told me, "One either does it or doesn't." So I guess trying is out of the question. Now, be aware that your dog is going to jump on you and push him away, growl at him or even throw him to the floor all the while growing, showing your teeth and flashing your eyes! If need be, throw a can of pennies down at the same time you are growling, stomping and going off on him. Remember this is an act.

Then almost in the same breath, smile and tell him to sit. Spend time with the calm dog that is sitting and pet him slowly. Don't get him excited. Don't pet him if he gets up. Growl at him again. This time tell him to sit. When he sits, pet him and smile. Count to twenty. He needs your attention when he is behaving. Guess you don't pay attention to him unless he is a pill. Stop that! If he continues jumping, get the leash on him and step on the leash so he can't jump, now tell him to sit.

Never pet or love a jumping dog, it is a very bad thing. It is against the rules. It is not to be tolerated. Why did you teach him that?

Sit and you get petted. Jump and you get killed. That, of course, is just another one of my stupid sayings in order to get a point across. (It is an act!) I never loose my temper. If I do its time to walk away. Learn the act and all will be well. (Just be sure you told your neighbors you are not harming you pup, nor are you really mad.)

Getting into the trash!

Getting into rotted trash is a survival instinctual trait!

It is also a learned behavior.

Know that this is what dogs do. Get that mind set. Then make sure you pay attention to that young pup you gave your house to! You may want to entice the pup to get into the trash in order to teach him never to get into the trash. Your dog should never be around or near the trash in the first place. My dogs are not even aloud to sniff the trash can. Not even look at it. If your dog ever does get in the trash, know that it is almost impossible to stop this behavior. You might have to just give up and put the trash up. Far away and out of reach.

My dogs never get into the trash. When I bring a dog into the house I watch him. I may remember that I put some chicken bones in the trash, or meat, or leftovers, or whatever. I may even put some in there if there wasn't any just to watch my dog. Then I would go out of the room and peek around the corner and wait. I might even stick a few hundred mousetraps around the trash and inside the trash. I would definitely make the experience something he would never do again. Then I would make sure I feed my dog well enough to have him not be so hungry that he would need to get into the trash. You know I give my dogs leftovers. I even give them chicken bones but I do not encourage you to do so.

The most important thing I can tell you about this subject is to not allow this experience to ever happen to your dog! If it has, sorry, Charley. You learn. You now have to continue to put the trash up and away.

By the way, if you get another dog while you have the dog that gets into the trash, your new dog will learn all the habits of the old dog. Now that you know that perhaps, you will protect yourself against the fact that the new dog will pick up these awful habits. Train them young. Set up the problems while you can correct them. Good Luck.

Teaching your dog to retrieve

This is absolutely very simple. Don't pet him unless he has something in his mouth! If you would like to read more on this, look for the book Training the Alsatian Shepalute.

Herding and Herd Protection Dogs

First let me say that all dogs were bred for one purpose or another and genetically they possess all the right stuff to do the jobs they were bred for. This is not abuse. Frankly in our opinion, its abuse to not use the dog for what it was bred for.

Because of this there are classifications that dogs are placed in according to what they were bred to do. Herding dogs and protection dogs are in the "working dog" or "herding dog" category. Alsatian Shepalutes are in the "companion dog" classification but can do other breeds jobs. They would not receive the high placing scores. They love to be with us and will walk the herds in, will protect and secure their territories and will learn anything you teach them.

Descriptions of:
HERD PROTECTION DOGS (Hpd)

Herd protection dogs are usually large dogs. Mean enough to take on the large predators such as bears, large cats, and large wolfs. Herd protection dogs stay out with the herd 24/7. That means all the time.

A large part of training these dogs relies on genetic breeding's the rest must be taught. The owner of a herd protection dog must be able to train the herd protection dogs to stay with the herd and must make the dog accept the herd as its pack. The herd protection dog has a territory that it calls home and will fight off any predators, man or animal who pose a threat to any of the livestock under its care. Not many dogs are capable of doing this job. The hardest part is staying and claiming the herd of livestock as its own. How does one get a dog to stay with the herd 24/7 with out the dog running away to the next farm, or with out the dog coming back to the house or front porch? How do you get this dog to stay out with the herd?

A herd protection dog is a rare animal to find. What dog in its right mind would attach itself to a flock of sheep? Or goats? Calling the herd its pack? (Far and few between). If the dog breed you intend to train for this job doesn't have the proper genetic combination for doing this, the dog would "come in". It wouldn't work.

Coming in means that the dog will leave the herd to be with, say other dogs, or your family. Companion dogs are dogs that prefer humans to other animals. So you see the Dilemma. A herd protection dog is not a companion dog; it's a working dog not to be pampered or the job taken lightly. It is a very serious undertaking and an extremely important job to any rancher. That Herd protection dog is vital to a rancher's income.

HERDING DOGS

Herding dogs are smaller and quicker, lighter on their feet than the herd protection dogs. They can not protect the herd from the large predators, though courageous, they may die trying. Their abilities lie in the stocking, running, crawling and mesmerizing of the herd animals in order to move herds as their owners wish. They are working dogs. A working dog has energy. Lots of get up and let's go! Far too much for me. Hehe. Herding dogs can be companion dogs. But to be truly happy, a dog that is bred to herd would be happiest herding or working. That includes running and letting off energy that was bred into that breed enable for that dog to do this job. To a rancher, sheepherder, or other livestock herder, this is a very serious job and vital to this persons income.

Ok, now that we know the difference lets compare the different breeds and their characteristics in handling both of these jobs.

Breeds of Herd protection dogs:

Alsatian Shepalute

This dog is as intelligent as the German shepherd but 55% less hyper. It is large like the Great Pyrenees weighing in at 80 to 100 pounds. I know of some Alsatian Shepalutes who weigh in at 130 pounds.

This is a friendly dog with a deep-throated bark. If not properly trained it may kill any animal that runs from it, as it is prey oriented. But, it is not so good at catching or cornering prey as this breed is kind of laid back.

The Alsatian Shepalute comes in all colors but timber wolf gray is the norm. Its coat is nicer than the Pyrenees yet is not as cottony. The Alsatian sheds its undercoat completely during the summer months so that it may bear the heat. Its ears are erect and can hear better than a drop eared dog. It is not a shy dog; on the contrary, it will approach or stand its ground.

This dog is easily trained as it learns quickly. If used for herd protection this breed of dog would need to be professionally trained at a young age to accept the herd as its family as it does tend to prefer human companionship. This dog will take on the big cats, wolves, and bears. I personally have no problem with any of the wild animals coming onto our farm. My dogs kill skunks and squirrels and though my neighbors have had large cats and bears on their properties, no large predators have come onto our land.

I secure the perimeter daily with our dogs and encourage them to urinate in all the corners of our property. Our dogs are professionally trained so much so that when attacking a small Shih Tzu, who stalked the chickens, the protection dog was called off while his jaws were around the little dog's body. The little dog was not harmed and wished to pursue the fight of the large Shepalute so I had to tell it to go home before it tried to bite my dog's hocks. We have chickens, guinea hens, turkeys and goats that roam our property freely.

Anatolian

This mastiff type dog from Turkey is white with light colors of tan and once in a while you may find a black one. I have found this dog to be shyer and not so mean as the Great Pyrenees. He normally patrols rather than herds. He has evolved a technique of very effective ambush. He is an active, rather long-legged dog with great agility and stamina and is hardy and capable of living under conditions which other dogs would find very trying. His coat is medium length, very dense, rather soft, and though pale colours such as fawn, with a black mask are most common, entirely black specimens do occur. In turkey his ears are cropped and he wears a heavy spiked collar for extra Protection. They are good dogs but shyer than the Alsatian Shepalute and lighter in weight. Not as easy to train or as social this in my opinion makes this dog an easier dog to train to stay with the herd. For some reason this dog actually thinks the flock or herd is its natural pack, though it does not try to breed with the herd. I do suggest neutering an animal that is solely used for herd protection to assure the dog would not wonder to the call.

Great Pyrenees

This dog is not usually hyper. Used with sheep, its work was that of guard rather than herder, and like the Anatolian Kara bash its main purpose was to protect the flocks against wolves. This resulted in a very large tremendously strong and fierce dog not at all suited to being a family pet if properly bred for service.

30 inches high at the shoulders. Double dew claws on hind legs. All white with hanging drop ears. He carries his tail low when standing and curls it like a plume over his back when he moves. At one time there were temperament problems, and some of these dogs were far from trustworthy. The breed clubs realizing this wrote into its standards that any "bad temperament should be considered a fault."

Having trained a few of these dogs, my personal opinion is that these dogs are mean and can not be trusted when it comes to children wandering on your property. If I owned one I would be careful of any strangers entering the pastures that they may get mauled to death. I would need a good fence and electrical warning signs even if the fence wasn't charged. They will take care of your large predators such as mountain cats, wolfs or coyotes even bears. These dogs will stay out with the herd easily as they rather prefer this. As a canine they will protect their territory.

German Shepherd

This is a dual-purpose dog. Both with brains and herding instincts. It is a working dog and requires lots of exercise. It is mean enough to protect any herd from any predators large or small. It has a terrific natural chase instinct and will kill anything that comes onto its property with a deadly quick snap of the jaws. It is a family companion dog with the ability to think situations out. The only fault this dog has is that it will kill chickens and may kill the newborn livestock, depending on the shepherd lines you get. Most of these dogs would not do the job as they once did many years ago. It's been to long. You may get a good line though and with excellent training you may have one that could do the job, but don't turn your back on the small livestock, you might find only bits of hair or feathers on the ground. Being more of a companion dog in the United States this breed will not stay out with the herd.

The following breeds are not recognized by A.K.C but realize A.K.C is only the American kennel club!

The following breeds are grouped together as guard dogs and hauling dogs throughout the world.

Ainu, Canaan Dog, Dogue de Bordeaux, Entlebuch Mountain Dog, Eskimo dog, Estrela Mountain dog, Eurasier, Fila Brasileiro, Great Swiss mountain dog, Greenland Dog, Hovawart, Sanshu, Landseer, Leonberger, Neapolitan Mastiff, Norbottenspets, Pinscher, Perro de Presa Mallorquin, Portuguese cattle dog, Portuguese shepherd dog, Pyrenean mastiff, Spanish mastiff, Tibetan mastiff, Tosa Fighting dog

Breeds of Herding dogs

Alsatian Shepalute

As a herding dog, this dog would not fair well in competition. Sorry, it is not quick enough and is too large. Our Shepalutes do go out to the herd and we walk the herd back. No running for us! Shepalutes stay beside there owners and protect both the herd and owner. If one shot off a gun or practiced the bullwhip while herding, the Shepalute would not be bothered. The Shepalute will "go out" and "go around" and "wait" and "down" on commands or signals. The Shepalutes will gather a leash in its mouth and bring an animal in.

Australian cattle dog (not A.K.C)

In the early days the Australian settlers were faced with the problem of controlling vast herds of cattle and flocks of sheep grazing over huge tracts of land. The only answer to this type of problem was to use dogs. The Australians started with a dog known as the Smithfield, but this was a large and somewhat cumbersome animal that was too noisy and upset the herds making them nervous.

A man named Timmins of New South Wales (a drover by trade) crossed the Smithfield with the native Dingo around 1830 and produced a litter of red, bob-tailed pups, which were known as "Timmins biters". In 1840 a landowner named Hall imported a couple of smooth blue merle collies from Scotland and these, in turn crossed with the Dingo and the Timmins dogs, produced what were known as "Hall Heelers".

From that time onwards there were continued efforts to improve the breed, and the present-day blue and red mottled cattle dogs are the result. A standard was drawn up in 1897 for what was then known as the Australian heeler and the name has since been changed to the Australian cattle Dog. This dog will work in the exhausting conditions of the Australian climate and rarely flags. He has a short coat.

Australian Kelpie (not A.K.C)

In 1870 a grazier named Allen imported into Australia a pair of so-called Fox collies from Scotland. They were black and tan with prick ears. They mated on board ship and whelped a litter of two, one red, and the other black and tan. They proved to be good workers fast, quit and capable of working over wide areas. At about the same time a smooth black and tan bitch named Kelpie appeared in New South Wales and was mated to a dog of the original litter which then produced a famous winning bitch also named Kelpie. She was the foundation of the breed as all her progeny turned out to be good workers with sheep, and more or less in her honor they were all called kelpies.

Australian shepherd

This is a smart dog. Easily trained but far too hyper for me. I couldn't keep up with its energy. This dog loves to work! This dog would protect its herd and property with its life. This dog is ready and willing to go on a drop of a dime. It would not fair well in a head on fight against large predators but it may out dodge them or out run them. I have seen amazing things with this dog. They can be taught anything a trainer can think of. They open and close gates and their herding instincts are always right on.

Beauceron (Berger de beauce)

This is a French sheep-herding dog. He looks something like a short-coated German shepherd dog though with rather more elegance. F.C.I registered. Used for guarding the flocks as well as working them, he has certain aggressiveness allied to his tractability.

Belgian sheepdog (Laekenois)

These dogs' origins are in the small area of Belgium near Antwerp where one of his duties is to guard the fields. Dual working. Fawn in color with some black on the muzzle. Long rough and shaggy coat with a certain hardness of texture, long on the body and shorter on the head. Twenty-five Inches at the shoulder with a long neck.

Belgian Tervuren

Once shown and bred as Belgian shepherd dogs. When herding, this dog was a person's food and income or livelihood, the only requirement for a sheep-herding dog was that he do the job and do it well. These dogs do the job well. Too hyper for a plain ol' companion dog in my opinion but remember this dog has to be hyper and have that drive of a great working and herding dog.

This dog is a square standing dog with long guard hairs and a thick undercoat. His color is from fawn to mahogany with a black overlay or double pigmented. He is thinner with a lively gait never seeming to tire.

Border collie

This dog was said to come from pedigrees of dogs that belonged to the international sheep herding society. He is a typical collie type strong and active. He comes in two types of coats, the rough coat, which is a medium length with a tough outer coat and a thick undercoat. The other type of coat is the smooth coated Border collie, which is shorter. These dogs are basically black and white but in their standards of the breed, which was approved in 1963, it states that many different colors are acceptable. I have seen these dogs come in many different sizes from very large to very small.

Collie

It seems that collies have been herding sheep for as far back as flocks of sheep have been on the earth. A lot of breeds have developed out of the original sheep herding collies. The collies of today can probably still herd sheep, but I wonder about that long show coat? They sure are a beautiful dog, one of the most beautiful that I know of! They do come in a short smooth coat, which is easy to care for. My personal opinion is that they bark a lot. But besides that they are a very smart and gentle dog. A large dog with more energy than the Alsatian Shepalute. They can jump very high for their size.

Corgi's

Here is a nice small herding dog. These dogs do not bark too much and they have the energy to herd anything, as do most herding breeds. They come in a variety of coats from long to semi curly, which they should not be to a short thick coat. They are stocky dogs, which slows them down some. There are two different breeds of corgis with two different standards. I believe that the corgis were the same dog and a mutation occurred when they had pups with bobtails so the divided the corgis into the cardigan with a tail and the Pembroke with the bob tail. He is a tuff little dog. A lot of folks use this dog to herd cattle as the herd is slower and the corgi has short legs.

Great Pyrenees

This is a guard dog for herds not really a herding dog though some countries do use this breed for herding. He is not the best herding dog as he lacks the energy.

German shepherd

The German shepherds can herd, but you have to watch out for this dog eating your profits. The German Shepherds I have known would of course have to be shown how to herd where other herding breeds do it naturally. I would never choose this dog if I wanted a herding dog. I would have to say that the Australian Shepherd is the best all around herder. If A.K.C had a herding dog section this dog would need to be placed in the working dog classification as he fits better in that category.

The "German" German shepherds could still herd with non stop attention on the herd, but the American German shepherds would tire. I would trust the American bred shepherd before the German bred shepherd if I needed a dog to not harm the livestock.

The following are other herding dogs that you may look up if you are interested. Lapenporokoera, Lapphund, Maremma sheepdog, Mudi, Berger Picard, Polish Tatra Herd dog, Portuguese Sheepdog, Pyrenean sheepdog, Pumi, Schapendoes, Swedish Vallhund.

Chapter 13

Dog Laws

The bulk of today's laws are rooted in the past. Judges make law, which tilts in the direction of previous law. Current law is based on the common law, which has roots in Greek and Roman law. Greek and Roman law stems in part from the Bible.

In the United States, the law is based on the constitution and the preservation of constitutional rights. Under the constitution you have certain obligations or duties, but you also have certain rights which are protected. (So far).

Some of the more important guaranteed rights are:

1. Freedom of speech and the press.
2. Freedom of assembly and religion.
3. The right to vote and to hold public office.
4. Equal protection under the laws.
5. Protection of individual privacy.
6. Full enjoyment of one's property.
7. Due process of law.

There are two basic principals under which the laws are supposed to operate:

1. Fairness — if a wrong has been committed, the remedy should as much as possible, make amends for the wrong.

2. Equality —- all individuals are to be treated alike.

I would also like to include a little bit on the **spirit of the law**, which has seemed to be disappearing while the letter of the law is being more and more interpreted and amended.

The spirit of the law is that for which the law was written. It is not and was not spelled out specifically to be picked apart and indiscriminately interpreted by persons who wish to make of it, as they will. The spirit of the law for me is the most important part of our legal system and that, which is being denied or over looked. It is so important that I feel that all people should study the law! Too many innocent and undeserving people will be put in jails in the future if this country continues on its current path. I hope and pray that others will be able to see these things and will help keep our citizens out of the jails we have in this country. Learn the laws!

When I went through the police academy, one of the many things that I learned was that a police officer could arrest anybody he wanted! What you say? How can that be possible? Believe me it is true and they know it! All they have to do is just follow a person around and/or wait long enough and that person would commit a crime or offense. How could that be? Because there are so many laws!

Arresting citizens brings more money into the legal system. It also protects law enforcement employment positions by guaranteeing jobs. Most of the public is unaware of the vast number of laws. More and more laws are being added everyday, slowly taking away our rights to live our lives as we see fit.

Animal Law History

In the beginning of mankind, humans would follow animals around and learn from them. We became scavengers ourselves and we ate what other animals killed.

As the world progressed and humans raised animals for food, scavenger dogs were chased away from the flocks and/or killed by stoning or spearing.

Rules governing rights of property owners began to appear and a collection of laws evolved concerning all living beings and their behavior towards one another.

Regulations became necessary because of the increase of the number of humans and animals and the likelihood of interference with the rights of others. These growing numbers of regulations brought in money to the state or county. (Regulations and laws are restrictions on what we and our animals are permitted to do.)

In large cities a pet owner cannot permit his animal to disturb (bark or soil property of another) without violating a law.

The offending neighbor is also restricted in his behavior. He is not permitted to discipline, harm or damage another's property (pet).

Many laws govern the dos and don'ts of animal control and individual liberties of both animals and humans.

Penalties to owners are fines and having their animals quarantined or locked up.

Unlicensed, unleashed dogs are considered a public nuisances and a pet owner may be fined because an animal regulatory law has been violated. His pet is confined and if the owner does not redeem his pet the dog may be put to death.

Animal Laws and Cruelty Laws

In the days of the presidency of Abraham Lincoln, Henry Bergh "The Great Meddler" was sent on a special presidential mission to learn about laws concerning cruelty to animals. (Actually the president just wanted to get rid of the pest!)

He couldn't find enough inhumanity towards animals here in the states so he had to look elsewhere. His travels took him to Moscow, Russia, England and many other countries. Henry's observations and information helped gird him with the necessary knowledge to be able to establish the guidelines for the American Society for the Prevention of Cruelty to Animals (ASPCA) which was located in New York City.

This organization had a charter and was under the laws of the state of New York. Early regulations were enacted covering the prevention of cruelty to animals.

One day, the case of an abused child was brought to the attention of the Meddler, who took a policeman and went to investigate. The abused youngster, by the name of Mary Ellen, had been beaten and battered by a foster mother who had custody of her.

The Meddler tried to have the woman arrested for cruel and inhumane treatment but was informed that there were no laws preventing cruelty to a child in the state of New York.

The Meddler found that there was not even a law in the United States that he could call upon to prosecute the foster mother.

At that time there had been a law against cruelty to animals in the state of New York and it had been there for approximately two years.

Henry the Meddler then got a warrant charging the woman with cruelty to an animal. The foster mother was arrested and brought into the courtroom confident that she would get off. Henry then brought the beaten and bruised child, concealed within a blanket, into the court room. Henry then told the court that the animal that was abused was a human child. The evidence was the bruises. The foster mother was then charged with cruelty to an animal and that case laid the foundation for the laws preventing the cruelty to children. ("Laws Pertaining to Animals", Less. 30, p.1)

Some Animal Laws

Privileged entry — For the purpose of discharging the duties imposed legally and to enforce the same, the pound master or any peace officer may enter on private property, except dwellings located thereon, as follows:

A. during daylight;

1. When in pursuit of any animal that he or she has reasonable or probable cause to believe is subject to impoundment pursuant to applicable law—-

2. To impound or place in isolation any animal thereon which he or she has any cause whatsoever to believe or suspect has rabies or is a biting animal.

3. To inspect or examine animals isolated thereon pursuant hereto or other applicable law——

4. To impound an animal pursuant to section —— (immediate threat to persons or property).

5. To seize an animal pursuant to section —- (failure to take action of an order within limited time).

B. During night time

1. When in pursuit of any animal which he or she has reasonable or probable cause to believe is subject to impoundment pursuant hereto or other applicable law.

2. To impound or place in isolation any animal thereon which he or she has any cause whatsoever to believe or suspect has rabies or is a biting animal.

3. To impound an animal pursuant to section —- (immediate threat to persons or property)

As a condition of the authority set forth in this section, except where time does not permit in an emergency or when in fresh pursuit, before entering upon private property a reasonable effort shall be made to locate the owner or possessor thereof to request permission to enter upon such property and to explain the purpose for such entry.

167.315. Animal abuse in the second degree.

(1) A person commits the crime of animal abuse in the second degree if, except as otherwise authorized by law, the person intentionally. Knowingly or recklessly causes physical injury to an animal.

(2) Any practice of good animal husbandry is not a violation of this section l.

(3) Animal abuse in the second degree is a class B misdemeanor.

167.320. Animal abuse in the first degree.

(1) A person commits the crime of animal abuse in the first degree if, except as otherwise authorized by law, the person intentionally, knowingly or recklessly:

(A) Causes serious physical injury to an animal" or

(b) Cruelly causes the death of an animal.

(2) Any practice of good animal husbandry is not a violation of this section.

(3) Animal abuse in the first degree is a class a misdemeanor.

167.322. Aggravated animal abuse in the first degree.

(1) A person commits the crime of aggravated animal abuse in the first degree if the person:

(a) Maliciously kills an animal; or

(b) Intentionally or knowingly tortures an animal.

(2) Aggravated animal abuse in the first degree is a Class C felony.

(3) As used in this section, "maliciously" means intentionally acting with a depravity of mind and reckless and wanton disregard of life.

167.325. Animal neglect in the second degree.

(1) A person commits the crime of animal neglect in the second degree if, except as otherwise authorized b law, the person intentionally, knowingly, recklessly or with criminal negligence fails to provide minimum care for any animal in such person's custody or control.

(2) Animal neglect in the second degree is a class B misdemeanor.

167.330. Animal neglect in the first degree.

(1) A person commits the crime of animal neglect in the fist degree if, except as otherwise authorized by law, the person intentionally, knowingly, recklessly or with criminal negligence:

(a) Fails to provide minimum care for an animal in such person's custody or control/ and

(b) Such failure to provide care results in serious physical injury or death to the animal.

(2) Animal neglect in the first degree is a class A misdemeanor.

167.335 Exemption from ORS 167.315 to 167.330.

Unless gross negligence can be shown, the provisions of ORS 167.315 to 167.330 shall not apply to:

(1) the treatment of livestock being transported by owner or common carrier;

(2) Animals involved in rodeos or similar exhibitions;

(3) Commercially grown poultry;

(4) Animals subject to good animal husbandry practices;

(5) The killing of livestock according to the provisions of ORS 603.065;

(6) Animals subject to good veterinary practices as described in ORS 686.030;

(7) Lawful fishing, hunting and trapping activities;

(8) Wildlife management practices under color of law; and

(9) Lawful scientific or agricultural research or teaching that involves the use of animals.

167.340. Animal abandonment.

(1) A person commits the crime of animal abandonment if the person intentionally, knowingly, recklessly or with criminal negligence leaves a domesticated animal at a location without providing for the animal's continued care.

(2) It is no defense to the crime defined in subsection (1) of this section that the defendant abandoned the animal at or near an animal shelter, veterinary clinic or other place of shelter if the defendant did not make reasonable arrangements for the care of the animal.

(3) Animal abandonment is a class C misdemeanor.

Cruelty to animals.

The definition of animal abuse is vague. Please go now to the glossary and read this for yourself so you will be aware of how the ordinary man reads the definition.

Housing Law

Housing law in NYC is such that if an animal has lived in an apartment building containing 3 or more units for 3 months or more and legal action has not been taken to evict the animal, it can stay, provided thee animal has not been deliberately hidden. Before anyone gives up a wanted pet due to housing problems, consultation with a knowledgeable attorney is advised. The laws in your state and county may differ. You can access your laws through your local library or on the web.

Animal Rights?

A legal right must be recognized by the law as a legal right! Humans can and do have legal rights. All non-human animals are things under the law. Things have no rights they are considered property so far.

Pets are Property —- A dog that lives in a junkyard may sleep under a truck or find shelter as all animals do. Shelter is defined as something that shelters one from the elements. This junkyard dog may drink as wild animals do and have been doing since the beginning of time, from a rain puddle or a stream it can get its water. There may be a large troth or pail in the yard that is filled with water. This junkyard dog may exist on food that is thrown out to him. This animal is, under the law, housed and fed. Unless this animal is sick, injured or the ribs are showing considerably, one cannot, under the law as it currently exists, tell another how to keep their pets. If and when others do get the legal ability to tell others exactly and precisely how they must feed and care for animals (and soon ones own children), then we will all be in trouble and possible be put in jail. We become a police state, a socialist government and enslave ourselves.

Estate housing for Pets

Pets can not inherit property. A pet owner who wishes to provide for pets (after that person's death or during an inability to care for the pets) must appoint someone to look after the pets.

Extreme animal rights groups (animal right extremists) are Homo sapiens who are out in the extreme left field of the humanitarian world. They place the welfare of animals above their own fellow human beings.

When asked this question:

"If there was a car accident in which there was a mom, dad (driver) 2 children (in the back seat) and a pet dog, in where the car swerved off the road and was about to go up in flames. And you were standing beside the road about 18 feet from where the car came to a complete stop. You were in no danger but the inhabitants of the car were all going to go up in smoke in less than 2 minutes. You could only run to the car, open the door and grab one being, whom would you choose?

All animal extremists will answer: The *dog*.

All animal right groups' place animal's rights above human rights because they feel they are the only ones speaking for the animals.

Proposed changes to the Animal Welfare Act

The animal welfare act and regulations can be accessed at: http://www.nal.usda.gov/awic/legislat/usdalegl.htm

The animal field alone generates multi-billions of dollars and is very much a webbed and integrated system of billionaires and in my opinion, "Mafia type" organizations that are all out for animal loving humans and their money, however they can get it! That's their job's to get un-taxable income or donations from the public.

Off shooting branches of these million dollar organizations are organizations that do not even physically exist where there address' claim them to be.

A spokesman on the TV news station last night gave a definition of a terrorist group, as "A terrorist group does not have a return address". Yes, so right! So in my opinion these folks are terrorists acting within the United States and being allowed to do so!

How about these non-profit Organizations recruiting animal lovers to picket and solicit for them while helping to close down pet shops, grooming shops, circus' and rodeos throughout the united states and all the while getting away with it! They even pay the unemployed to hold the signs. These sign holders don't even know why they were holding the signs. They told me they were getting paid to do it!

These invisible groups may even change their names several times so that they can not be located. They use bogus addresses in case they spew their lies about private citizens like you and I and bankrupt them leaving their families homeless and destitute! These small off shooting organizations of 'do good' individuals leave the door open for those multi-conglomerates to take over, and they have! Look around your town. Do you remember when there were old barns converted into pet shops where you could buy pigeons, baby chicks, pigs, goats, etc.? How many do you see now? The folks that owned those shops were small business owners (just like you). They paid their rent, fed and clothed their children and sent them to school all in the pursuit of the American dream.

Animal Right Organizations
Humane Society or HSUS (Non-Profit organization)

Police officers, the humane society and your local pound work hand in hand, one taking care of the other!

Some of their web sites are:

http://www.hsus.org/ace/article-printer-friendly?contentid=11568

I'm sure there are thousands more....

The Humane Society of the U.S was incorporated in 1954,
Delaware Tax Status: 501 (c) (3) and is tax exempt.
Staff Size: 210 employees
1995 Income: $38,102,167
Donors: over 2 million
Board Size: 23 members
Principals:
John A. Hoyt, Chief Executive Officer
Paul G. Irwin, President

HSUS pays Hoyt, $237,871 and Irwin $209,051 in salary per year. HSUS bought Hoyt's home in German town, Maryland for $310, 00 in 1986 and allowed him to live in it until 1992 when Hoyt bought a home in Virginia and HSUS sold the Maryland property for 351,500.

HSUS paid Irwin $85,000 for renovations to a Maine cabin Irwin held in trust for HSUS and used by Irwin and his family for vacations. HSUS sold the cabin for $98,237 in 1995.

In 1988 an internal investigation by the board of directors of HSUS revealed "salary supplements" of $41,000 for Hoyt and #33,000 for Irwin over a three year period.

In 1982 Hoyt received a $1000, 000 interest-free loan from an HSUS board member. Another board member subsidized overseas travel for Hoyt's wife for years.

Legal documents showed that Irwin collected $15,000 in executor's fees from the estate of an HSUS board member, without notifying the board of directors in advance, as mandated by the HSUS code of ethics.

Irwin owns five houses, including a $786,500 residence in Darnestown, Maryland as well as a Mercedes, Lincoln Town Car and a Corvette.

Affiliates:

Humane Society International,

Earthkind USA,

The Wildlife Land Trust,

Earthkind International,

International center of Earth concerns,

National Humane Education Center,

Center for respect of Life and Environment,

National Association for Humane Education and Environment.

The Humane society of the United States (HSUS) raises enough money every year to be able to completely run one animal shelter in every state with enough money left over to spay, neuter, feed and save the lives of thousands of homeless cats and dogs!

The rescuing of stray cats and dogs and other animals is done by many different organizations and volunteers who really believe that it takes more than just the humane society to do all this work. There are loads of humane societies throughout the U.S with many different names.

Each and every year this tax exempt society with the words "United States" written within it, capitalizes on our concern for the lives of all the many thousands of homeless animals that they claim need our money.

The following are some true stories involving the humane society. I am sure there are a lot more out there.

1. First true story:

Around 1976 a person who loved animals took animals off the street to find their owners. She spent her own time and money not the governments (yours). While training a small dog, she tied this dog to a tree right outside the front door where she could keep an eye on him. Training this dog not to wrap himself around the tree, but to go in the opposite direction of the pull isn't too difficult, and can be done. (As a trainer this is one of the things I teach my dogs. The command I use is "varousse", or "go-round").

This little dog had wrapped itself around the tree several times and the dog was learning fine. When the humane society got to this woman's house the dog was wrapped around the tree so that the dog had about a foot of rope in which to lie down and wait for the next lesson. Of course the humane officers wouldn't hear of such a thing as training a dog to unwrap itself!

This same lady had 2 pups about 3 months old of a mixed lab setter type. Plus she had an older shepherd in the bedroom. All these dogs were living in a 2-bedroom condominium with a patio, grass and a wooden 6-foot fence.

The owner of the complex had called the humane society and the officers had come to write the dog lady up for all they could find her in violation of. That was their jobs. They swore to uphold the rights of animals! The officer had first asked about the little dog tied up on the leash outside the front door. The dog lady said it was not her dog and that she was training it for a friend. Well the officer took the little dog and put it in the truck. Then the officer asked if she could come in and was shown to the back yard. She saw that there was no doghouse and that a large bowl had no water in it (she didn't see the water lick). She wrote the dog lady up for not having 2-dog house's and no water in the bowl plus there was fecal matter in the yard. The officer then left giving the woman a citation to fix the problems.

A week later when the humane officer came back she was still not satisfied. The dog owner had built a large doghouse and the dogs had the same water lick on the outside faucet as they had before but the officer still had not seen it. Three humane officers had shown up and a police officer was there to make sure nothing was going to happen when the humane officer tried to take the dogs. The dog lady was hysterical and called the police station complaining to the chief. She then called a lawyer hoping to get some help. This can't be happening, not in America! She brought these dogs into the world!

All the humane officers had guns and badges and they had asked where the large shepherd was. The dog lady told them the shepherd was in the bedroom but that they would have to shoot the dog because it was protection trained and they did not have her permission to go into her bedroom nor to take any of her dogs. Permission not needed! Because the officer didn't see the water lick and because there was only one dog house, both of the 2 lab mix pups who were under 6 months old were taken to the humane society in Oxnard.

The humane society told the 22-year-old dog owner and her father that it would cost them to retrieve the animals at 80.00 per day. The dog owner's father then took pictures and taped the conversations with the head honcho in charge down there. The father stated "you mean to tell me that if we paid you folks $80.00 per day, then you would release the dogs to us?" The father was furious, "that's like a police officer giving me a ticket and then saying if you give me some money I would drop the charges. "No way" he said "I want to go to court!"

After 3 months there was still no word as to when the dogs would be returned and both pups had contracted kennel cough. Since they missed their training period they were too hyper and kennel crazy for any person to want.

Finally after months of trying to find out when the court date was scheduled, the district attorney dropped the case and the dog lady was called to come retrieve her pups at no cost to her.

2. Second true story:

When a grooming shop/school opened and the groomer displayed customer charges on a bulletin board for teeth and ear cleanings (along with nail clippings, wing clippings, bird baths and nutritional information) seven officials from the humane society, county pound, and the cities code enforcement department visited this shop. They carried guns, badges, clipboards and a camera. They bombarded the little shop to interrogate and record the goings on. (Actually what really happened was that these folks didn't have the backbone to go to the owner of the store and talk with her about what was on their minds). Since the code enforcement officer couldn't find any records of inspection of the premises (which happens to be a routine job that all new businesses go through before opening day)then the whole bunch of them meandered through the shop, without the owners consent, to try and find anything to write her up on. Nothing came of it. All conditions were appropriate and no charges could be filed.

3. Third true story:

When a ladies Schutzhund III German shepherd (worth over 1 thousand dollars to her) traveled over 20 miles (following his owners van) didn't come home, the missing dog's owner called the humane society and asked if such a dog was found. This dog lived in the national forest and did not have his collar or identification tag on.

The humane society secretary told the owner that it appeared that her dog was there. (This was before the secretary knew the identity of the owner). So the owner closed her business to travel 23 miles into Ojai to retrieve her beloved pet. When she got there the receptionist was busy being nicely rude to a man who had a litter of kittens. The owner of the German shepherd befriended the owner of the litter of kittens and asked him if he didn't realize that these folks really didn't want his kittens and that the receptionist was trying in a nice way to belittle him for allowing his cat to have kittens. The man then realized the abuse and rudeness that he had encountered and left the building. Meanwhile the receptionist had left the front office and was now returning with the top brass to discuss the German shepherd in question.

The owner of the German shepherd was told to go into the back and check the kennels. Though she didn't know where to go, she was not helped. The owner of the shepherd saw her dog and the dog saw her and sat politely to be retrieved. The owner of the shepherd asked if she could get her dog and was told that she had to correctly identify him. The owner asked "how would you like me to do that? This is my dog. His name is Omo von Der Dodo. He is a registered Schutzhund three German shepherd from Germany who speaks German and he is my dog." (The rude receptionist laughed at the fact that a dog could speak German.) "If you would allow him out of his kennel I could put him through his paces (speaking to him in German) and you would see that he is my dog." (Come on how many German speaking schutzhund three dogs do they get a year?)

They denied the Shepherd owner her dog and said they needed proof.

213

The Shepherd owner then stormed out of the place. Drove thirty minutes back to her place of business, got the record book with photographs and awards and traveled another 30 minutes back to the humane society only to find out that they had just took her dog to the county pound! The truck was just then pulling out of the yard!

The shepherd owner had to go the county pound, which was what the humane society wanted her to do in the first place, in order for her to retrieve her dog.

When she got to the pound she was told that the humane office had contacted them about this case. (Proof of cahoots.) Then they required the owner of the shepherd to pay a fee for a county license! The dog had a city license, but now had to have a county license. If the shepherd owner would have had a county license then she wouldn't have needed a city license. That's what she was told.

These are all true cases. I know for a fact because they happened to me.

I am absolutely sure that there are thousands upon thousands of stories just like these and I feel they all need to come to the attention of the public! Perhaps then we could fight back if we all rose up together and stopped allowing them to push us around.

Have you tried to adopt a dog from the humane society lately? My daughter called just today as I am writing this. She lives in San Diego and her friend with 7 children tried to adopt a dog. They told him that he had too many children. Boy was he mad. I was told that he was going to call back and see if they killed the pup yet! I don't believe they kill the dogs though, I believe they send them over to the county pound if they can't get anyone to adopt them. In our county I was informed that they keep dogs until they do find a home for them. And I had to ask her several times what happens if you do not find a home for them? She finally told me that if a dog is un-adoptable then they turn the dog over to the local pound. So the local pound is the one left holding the bag. This way the humane society isn't perceived the bad guy. That means the humane society can also say that they don't destroy animals.

By the way the humane society continues to use its non-profitable moneys that you donate to lobby their elected officials to push their beliefs on the folks that are giving them the money! I also heard a rumor that the SPCA and the human society were thinking about merging together. They might as well in my opinion!

If you go to the humane societies web site you will see where they are calling for thousands of animal group's to lobby to the fullest extent the law allows! They also spend a lot of their non profit money on all their lawyers to direct them as to how much each volunteer may lobby in able to stay within the law. Hmm? What do you suppose they are lobbying for? How many abused and neglected animals are there in the United States? And then again who is to say what constitutes abuse and neglect? It's a good thing we have a court system or we'd all be in jail!

Perhaps if I had better experiences with these organizations then I wouldn't be so adamantly opposed to the way they treat their supposed brothers and sisters in Christ while upon this earth.

I must tell you though that while I have removed myself from the terrible state of California I have encounter a warmer and friendlier group of humane society personal thus far here in Oregon. I hope they will continue to present themselves with politeness, friendly phone manners, and helpful information, even after they read this book!

Local pound or Animal Shelter —- The terms "animal control Department" and "Department of Animal control" mean the animal control section of the Environmental Health Division of the Environmental Resource Agency of the County of (you fill in). The term "animal pound" means any dog pound, animal shelter, and temporary animal pound or pound vehicle owned or operated by the county of _____.

These guys have trouble being rude because they are on the state or county payroll and supposedly work for you and me. Heck, millions of our tax dollar's go to pay their bills and feed their kids. Most of these folks are pretty nice and polite about telling a person off, especially if you don't think the same way they do. (And don't you know that they love animals more than the public does?) They even put out signs to remind you that they love animals. Have you seen them? Big banners in front of the local pound?

These folks do try to educate the nasty public who abuse and miss treat animals! That's where all their profits are going. Check it out in your town and I will guarantee you that you will be absolutely <u>astonished</u> by the total amount of money they claim they use to educate the public! Somebody did just that in the city of Oxnard and they were so surprised at the out come that they confronted the city! Boy did I stand up and cheer when I read that in the local paper! Thousands of dollars spent in "educating the public". That's where <u>they said the money was going</u>! I never got any education from the county pound, did you? (I never got much education from the government school system either.)

Again, these folks work for the county and have guidelines so they can be held responsible for their rudeness. Most of them are very nice about letting you know that they are more compassionate than you are and that you don't know much about how to raise, train, or feed your animals. I have personally stood there and watched county pound officer's miss inform the public. The incident was that a family wanted to adopt this little Doberman looking dog. It happened to be a mini-pin (miniature pincher). Well the officer told the family that this dog would grow up to be a large dog. The family moved on to the next pen.

Another thing I should mention about your local pound or animal shelter is that each state has written laws that govern these folks so these guys are doing one heck of a job for all they have to put up with. Any job that requires public relations, or dealing with the public has to be handled tactfully. Also, Lots of folks mistake these guys for those animal right group's who they are not too far behind. Some of the persons who work for this government agency also volunteer their time to those other organizations and may also believe strongly in the animal rights agenda.

Here is a little bit of law that may be the same in your city.

Center for Animal Care and Control (CACC)—- In NYC, the Rescue and shelter duties of the ASPCA have been turned over to the center for Animal Care an Control (CACC), 326 East 110th street, New York, NY 10029, telephone (212) 722-36320. CACC was created in January 1995 by the City of New York. It is regulated by the NYC Department of Health, and is funded in part by New York City. CACC was represented at the conference, and cited statistics of how well it was doing. Other representatives questioned the statistics. It seems much too soon to judge how well they are doing. Why do you suppose this Center was made up? Why is it city organized? Hmm.

Society for the Prevention of Cruelty to Animals (S.P.C.A)

— *The* ASPCA is under siege by a number of groups. You do know that they can just re-group and regroup under a different name. The folks who work in the SPCA will always find similar groups to work for. They are all the same folks.

These guys make a lot of money off the uneducated public! Especially the very sensitive animal lovers in the public who really want to believe these guys are really helping to stop the abuse.

Do you believe in nonprofit organizations? How about the Red Cross? Do you believe what the panhandlers tell you? There is a connection in my opinion.

I really wouldn't mind giving money to cause's like these except that these folks cheat the public by hiding what they really stand for. And they just out and out lie!

Remember when the Red Cross said they would give your donations to the suffering families of 9-11? If you had listened carefully you would have heard it differently. What I specifically heard them say in the television Interviews was that "the money would go to help disaster related families". The public just got it in their minds it was for this specific disaster. The public wanted the donated monies to go to this disaster and only this one. The Red Cross of course had to give in to all those thousands of complaints so they wouldn't get into a lot of trouble with the public who donates all that hard earned money that pays their bills. The Red Cross of course fired the head honcho at the time to make it look as if they were handling things. It really wasn't her fault at all and it probably ruined her life! I want to give her my condolences and to tell her that I understand. If I saw it, then I am sure others saw it too. We aren't all idiots.

Yes, the humane society and others like them make you believe you are helping poor animals with their advertising lies when all you are really doing is giving them money to support their organizations and other organizations like them. You are giving them money so they can harass good people with their macho attitudes. And of course no one knows how to love and care for animals but these folks and others like them. Have you gotten their advertising in the mail of the same dog on a chain with his ribs showing and his head down? That dog wasn't even in the U.S.A.!

I refuse to have any dealings with liars, cheaters and folks who put words into other people's mouths and then turn what you say into what will help get more money into their organization's or shut down your place of business. The bible states to go away from folks like this. That one can not talk to such folks.

I commend a certain Hollywood actress who shunned these folks and stated the truths about them that they were putting words of lies into <u>her</u> mouth! Her agent responded by telling the SPCA that she did not support any organization that distorted her words. You go girl! Of course when I heard that I got up and danced! I was also surprised to hear that from a news reporter.

People for the Ethical Treatment of Animal's (PETA)

These folks are the creams of the crops! **Have you ever read what they stand for?** What they believe! They do not believe any animal should be "owned".

Do you realize what that means legally?

What that means is that the government and any organization like PETA could tell you what you could do with animals and what you couldn't. How you can use them and how you can't! Did you ever hear of socialism?

Fact: No true PETA member owns a dog, cat, bird or any other animal.

They could not stand to put fish in a fish tank and not have that fish be able to live its life outside its natural habitat.

Now I am not saying they are wrong to believe that. They have a right to believe in that and to treat THEIR animals how they believe they should. But don't tell me I can not eat meat or own a pet!

Watch their faces when the media asks them how many unwanted animals they have in their homes as their spokesman tries to get the public to adopt unwanted pets. (I've seen this) The media smiles as they ask the question thinking that this guy from L.A is going to say (with a smile), "too many!" But the spokes person from PETA gets this look on his face like he's caught, then he throws the public of guard by changing the subject. (He is a professional!) A look of puzzlement comes across the face of the news reporter for just a split second, as she realizes the truth, while this PETA correspondent continues his angelic behavior. This guy has practiced and practiced deceiving folks on a daily base. He is a pro at it.

These folks believe that all animals should be free to roam the earth and that we humans are the intruders. We should return the animals to the earth that really belongs to them anyhow. Believe me when I say that they would not mind it if all humans would disintegrate and leave the earth to the animals!

They believe the government (your paycheck) should put out the bucks to protect all animals from humans, even <u>your animal's</u> need protecting from you!

PETA use's money that you send to them to pass unconstitutional laws to move towards their goals. They are getting away with it because the voters, (Mr. and Mrs. Public and those of you who do not vote) don't realize what PETA really stands for! By the way, their goals can never be obtained. Want to know why? Because if their goals could be obtained it would mean the end of their organization! No one ever said these folks were stupid.

<u>Here's what they believe in</u> —- Animals and humans are equal (based on eastern religious philosophy). They believe that no animals should be used by humans for any reason, including pet ownership, which they call (exploitation). You thought it meant something different didn't you?

All Animal rights advocates adhere to the following agenda:

1. Abolish by law all animal research.

2. Outlaw the use of animals for cosmetic and product testing, and classroom demonstration.

219

3. Vegetarian meals should be made available at all public institutions, including schools.

4. Eliminate all animal agriculture (resulting in no chicken, fish, or meat for food, no leather for shoes or clothing).

5. Eliminate all herbicides, pesticides or other agricultural chemicals. Outlaw predator control.

6. Transfer enforcement of animal welfare legislation away from the department of agriculture.

7. Eliminate fur ranching and the use of furs.

8. Prohibit hunting, trapping and fishing.

9. End the international trade in wildlife goods.

10. Stop any further breeding of companion animals, including purebred dogs and cats. State and municipal governments should subsidize spaying and neutering. Abolish commerce in animals for the pet trade (resulting in no more pets).

11. End the use of animals in entertainment and sports (resulting in no more horse shows, dog shows, and animal actors.

12. Prohibit the genetic manipulation of the species (resulting in the elimination of critical medical research relating to cancer and other life threatening diseases).

These are the folks that are desperately trying to change the laws and take away the rights our forefather's gave their lives to protect! Laws that allow special armed personal who love animals to be able to go into your private homes and get your animals if your animal is suffering!

And what do you think suffering means?

Is it described in the law books exactly what suffering is?

Nope, it's left up to this organization's interpretation of the law, which is, all animals are suffering if they are owned!

As of this writing most of your pet stores are owned by the conglomerates and millionaires. Anyone owning or operating a small pet shop or grooming shop that breeds or sells animals is targeted by these animal right group's who will picket privately owned shops until they close them down! Anyone looking to open a pet shop must pass a hundred different codes written in lawyer language. This leaves the millionaires and Mafia with all their well paid lawyers in charge of <u>all</u> pet shops and the only place you can purchase pets is from the newspaper or private parties. It is getting harder and more expensive than ever to own and enjoy any pet. Look around, where do you get your dogs food? Do you know what that pet store supports? I personally support small business owners because giving my money to the large pet shops puts more money in the pockets of these animal right group's.

Legal status

In the past animals were classified into groups according to their legal status. Many of our laws stem from the old English common law. Livestock was a property status of great value and dogs were owned just for pleasure so they were not entitled to as much protection as the food animals like cattle or sheep.

Some of the old legal classifications of animals divided these pet animals into groups such as:

1. Dangerous 2.ferocious 3.mischievous 4.nuisance and 5.harmless.

Any animal could be transferred from the harmless category into the other categories if their behavior caused someone inconvenience.

The transfer of dogs from harmless to the harmful status was often because the dog had begun chasing livestock, which of course was of value. The owner of a livestock animal was then entitled to ask for redress.

Some of the earliest taxes on dogs were levied for damages caused to livestock owners.

Today the principle of dog licensing is based upon the premise that fees collected for licensing are used to pay for regulations and the enforcement expense of laws involving dogs.

Owner Liability

Some folks cannot handle the breed, which they wish to own. I have seen this over and over again. Many times the public would come to me to discuss a breed of animal that they would like to purchase and ask my opinion. The first thing that I would do was to give them a large book with the many different breeds in it so that they could make a list of which dogs they thought might be the one for them. Ninety-five times out of one hundred those people would continue to choose an animal on its looks alone. The look that appealed to them. And ninety-nine times out of one hundred it would be the wrong breed for that particular family. We would go over the words that described the breed and how the written words might hide the fact that this breed is particularly noisy, or dominant in its behavior or hyper. Have you ever noticed that the breed books never say anything negative about the breed?

(Did you ever have to make out any of those Navy personal evaluations? Every word in it is nice but it all means something different to the folks who understand the words. Well, that's how you have to read those breed books).

Then we would get back to reality as to the fact that no one in this family who is looking for a puppy has ever had a pet dog before. This drastically eliminates the choices for the breed whose personality and character will fit into their lifestyle. The new owners must understand that their choice of pet is their responsibility.

It is my firm opinion that all pet owners should be held legally accountable as to their pick of the breed of dog they choose to be a part of their household. I also believe that the seller or breeder should hold some form of responsibility as to the temperament of their dogs and/or choice of breed being sold to the public. As a breeder/seller I personally knew if an individual would not be able to handle the type of breed that they wanted and I would tell them up front. I would also require them to sign a form that would not hold me responsible. And of course I would tell them over and over again about the breed and how that particular breed might be too much for them to handle.

Anyone who owns a dog is legally responsible for controlling him. Most states make the owner liable for any personal injury or property damage caused by his pet.

Property Damage and/or Personal Injury —- Owners of animals who damage another's property or injure anyone may have to pay all the medical expenses. The victim may also be entitled to loss of earnings as well as any pain and suffering, scars or deformation that the victim will have to endure for the rest of his life. This can add up to millions of dollars!

Spouses may also be entitled to some money or income just because you could not control your animal. All this doesn't even include a malicious or reckless owner. A court may double or even triple the damage award for an owner of an animal whose conduct was particularly disgusting.

Who owns the dog? This could be proven with evidence as to who paid the veterinarian bills as well as with witnesses that saw the dog at your house for over a period of thirty days.

Dog Bites —- Most states have statutes that impose absolute liability on the owner for any injury or damage caused by their pet no matter who is at fault.

An injured person often sues under this dog bite statute because liability is automatic. This means that the victim does not have to prove that the owner did anything wrong. Strict liability may be imposed regardless of whether or not the owner was careless, tried to prevent an injury, or was unaware that the dog was dangerous.

In most states the burden of proof lies on the owner of the dog to prove that he was not at fault.

Postal Workers —- You must be sure that your dog does not interrupt the postal person's job in any way. You can be sued if a postal employee is injured by your dog and the postal service can pursue collections for all damages, including lost wages, medical costs, and damage to personal property and clothing plus any scars left for life along with any pain and suffering.

Negligence —- Habitually neglecting to do what ought to be done. Conduct that falls below a reasonable standard, the standard of care that the law considers you owe to others. If the injured party can show that you were unreasonably careless in controlling your dog, compensation may be awarded for harm that was reasonably foreseeable as a result. Not having a leash on your dog is a negligent act.

Nuisance —- Anything which repeatedly causes a substantial and unreasonable or unlawful annoyance, disturbance, inconvenience or damage to another. Some ordinances require two or more households in order to declare the problem as a nuisance.

Here are some examples of what a nuisance may be:

1. When your dog is barking all the time or late at night or after curfew when reasonable working persons go to bed, or if your animal utters barks or cries which are so load, so frequent and this continues over such a long period of time as to deprive other persons residing in two or more residences in the neighborhood of the comfortable enjoyment of their homes.

2. An unlicensed dog. Can you believe this one? I can't figure out why my unlicensed dog would be a nuisance to my neighbor! Some folks have told me that licensing your dog help to identify the dog and return it to its owner. During my lifetime as a child we had a few dogs that would get out and run at large and they were licensed. We called the local pound everyday looking for our dog and we were told that he was not there. Well, day seven rolled around and we all know that day seven in our pound was the day of death. Somehow it always happened that on day seven we would get a call from the county pound that our dog was there. We would go and get him of course. Well, being curious as I am, I asked the officer at the desk to look on the dog's file and tell me when the dog was picked up? Well I am sure you all know that our dog was picked up on the day we lost it!

3. A dog is a nuisance if he is off the owner's property, in the road or in any other public property or on somebody else's property where he doesn't belong.

4. Your dog is a nuisance if he gets off your property and inflicts physical injury upon any other animal or owner of that animal and it was unprovoked.

5. Any unprovoked or threatening behavior towards any person when such a person is conducting himself or herself lawfully and which occurs in such circumstances as to cause such a person to fear for his or her physical safety.

6. When your animal damages the real or personal property of another person other than the owner or keeper and this happens off the property of the owner or keeper.

7. When your animal dumps over trashcans or spreads trash of the property of the owner or keeper of the animal.

8. When your animal chases pedestrians, vehicles or ridden horses which occurs off the property of the owner or keeper.

Such public nuisance may be "abated" in accordance with the procedures set forth in sections.

There is that word - "abated" now that's not a bad word is it?

By the way, I suggest that you read the laws well as your dog may be "abated" (destroyed). They (the legal system) use an indiscriminant word, but it means the same thing as KILLED.

Photo Album

In Memory of Packer von der Schwarz

In memory of Sophie of Baloo 1994

Sit stay 5 weeks old! 2002

In Loving Memory of Rafael Petrowski

Photo Album
Every Picture tells a story, Don't it?

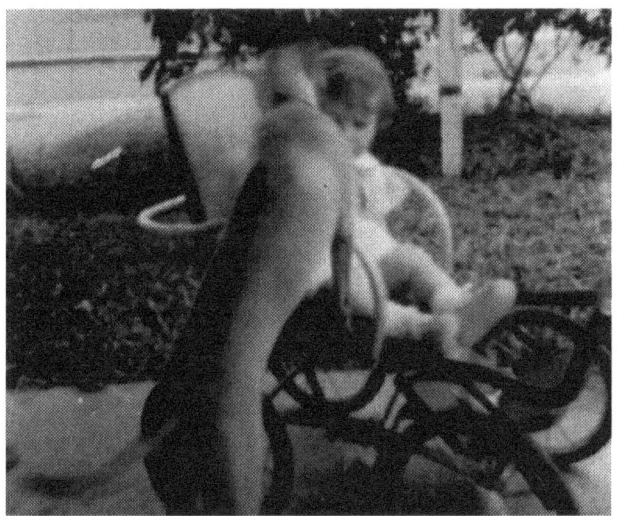

Lois and pepito 1954

Lois and pepito

Lois and brother brett in panama 1957

Louis in panama aviary 1959

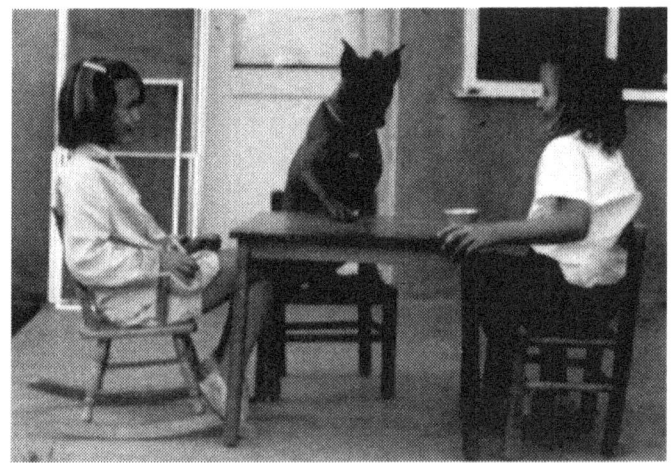

Robin Devil and Lois Oxnard 1962

Lois and Devilar 1963

Lois in Oxnard 1965

Lois in horse camp Girl Scouts camp Tequia

Bozeman Barnyard kennel Stratford, Ca.

Dog Shop Hueneme, Ca. 1978

Security officer for over 6 yrs

Lois Police Cadet Graduation

Lois graduation Truck Driving School San Diego, Ca.

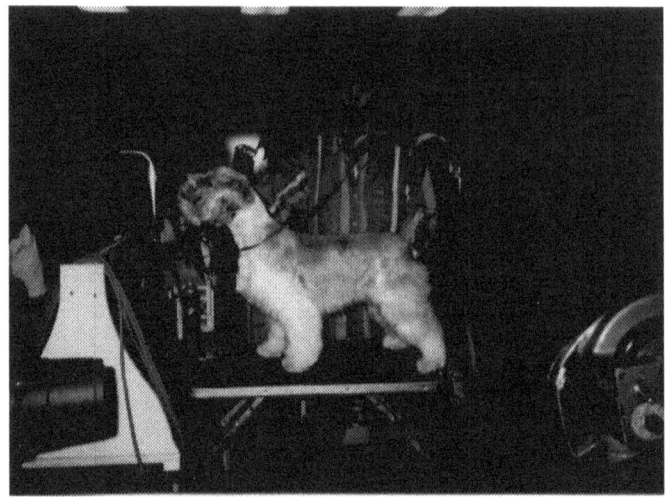

Lois second grooming shop 1985

third generation and sixth generation

Lois showing Sib the Siberian husky

Lois 1995

Jim and Trixy 2003

Alsatian Shepalutes 2003

Glossary

A.K.C —- the American Kennel Club

Albino —- an animal having a congenital deficiency of pigment in the skin, hair and eyes.

Almond eyes —— The eye set in surrounding tissue of almond shape.

American Kennel Club —- A group of individuals who put on shows and trails for all breeds of dogs that the club registers as pure of breed within the United States. It formulates and enforces rules under which dog shows and other canine activities in the U.S. are conducted. Its address is 51 Madison Ave., New York, N.Y. 10010.

Angulation —- the angles of the bone structure at the joints.

Anus —— The posterior opening of the alimentary canal through which the feces are discharged.

Artificial insemination —— The introduction of a dog's semen into a bitch's reproductive tract by artificial means.

Acquired characteristics —— It was believed that what characteristics one acquired because of life itself, was inherited by its offspring. This of course is not so.

Balance —- A nice adjustment of the parts one to another for the whole symmetry of the organism.

Bird dog —- A sporting dog trained to hunt birds.

Bitch —- Female dog

Breed —- Purebred dogs more or less uniform in size and structure, as produced and maintained by man.

Breeder —- A person who breeds dogs.

Brood bitch —- A female dog used for breeding. Brood matron.

Canine —- Any animal of the family Canidae, including dogs, wolves, jackals, and foxes.

Castrate —- To remove the testicles of the male dog.

Cat foot —- Round, compact foot, with well-arched toes, tightly bunched or close-cupped.

Character —- A combination of points of appearance, behavior, and disposition contributing to the whole dog and distinctive of the individual dog or of its particular breed. Expression, individuality, as considered typical of a breed.

Coat —- The dog's hair covering. Most breeds possess two coats, and outer coat and an undercoat.

Conformation —- The form and structure make and shape. Arrangement of the parts in conformance with the breed standards.

Cow-hocks —- Hocks turned inward and converging like the presumed hocks of a cow.

Crossbred —- A dog whose sire and dam are representatives of two different breeds.

Cryptorchid —- A male animal in which the testicles are not externally apparent, having failed to descend normally. Bilateral cryptorchidism involves both testicles not descending into the scrotum. Retained in the abdominal cavity. Unilateral cryptorchidism involves one testicle not dropping.

Cruel —- Disposed to inflict pain. Insensate or vindictive manner. Pleased by hurting others. Sadistic. Devoid of kindness. Given to killing and mangling or to tormenting prey. Bitterly conducted. Causing grief or pain. Unrelieved by leniency or softness. Extremely painful.

Cruel and unusual punishment —- punishment to include torture. Barbarous punishments, degrading punishments not known to the common law. Punishments so disproportionate to the offense as to shock the general moral senses.

Cynology —- The study of the canines.

Dam —- The female bitch. Canine.

Dewclaws —- Additional toes on the inside of the leg above the foot.

Dewlap —- The pendulous fold of skin under the neck.

Dog —- A male dog. Also used collectively to represent both the female and the male canine.

Dog show —- A competitive exhibition for canines at which the dogs are judged according to the standards for that particular breed.

Drive—- When this word is used in conformation and showing it means: A solid thrusting of the hindquarters, denoting sound locomotion. When this word is used in accordance to a working dogs ability: the hyper ness or want to do his job. It is inherited.

Featherings —- Longer fringe of hair on ears, legs, tail, or body. In most incidences the body hair is short and these featherings are approximately 3-4 inches long. They maybe longer.
Femur —- The heavy bone of the true thigh
Flews —- The chops, pendulous lateral parts of the upper lips.
Forearm —- The part of the front leg between the elbow and pastern.

Gait —- The pattern of footsteps at various rates of speed or a rhythm. In a perfect gait the top line of a dogs back would be straight and not roll, but glide.
Genealogy —- Recorded family descent
Guard hairs —- the longer, stiffer hairs, which grow through the undercoat and normally, conceal it.
Gun shy —- When the dog fears the sight or sound of a gun.
Handler —- A person who handles a dog in the show ring or at a field trail.

Hare foot —- Faulty foot. A long narrow and close toed foot, like that of the hare or rabbit.
Heat —- seasonal period of the female. Estrum.
Heel —- a command to the dog to keep close beside the handler's heel.
Height —- The vertical distance from withers at top of shoulder blades to the floor.
Hock —- The lower joint in the hind leg, corresponding to the human ankle. The part of the hind leg. From the hock joint to the foot.

Inbreeding —- The mating of closely related dogs of the same standard breed.
Interbreeding —- The breeding together of dogs of different breeds.

Lead —- A strap cord or chain attached to the collar or harness for the purpose of restraining or leading an animal.

Line breeding—- The mating of related dogs of the same standard breed, within the line or family, to a common ancestor. A dog bred to his grand dam or a bitch to her grandsire.

Litter —- The puppy or puppies of one whelping.

Malice —- A wish to vex, annoy, or injure another person, or intent to do a wrongful act, established either by proof or presumption.

Monorchid —- A male animal having just one testicle in the scrotum.

Mute —- silent. To trail without baying or barking.

Muzzle —- The part of the face in front of the eyes. The nasal bone, nostrils and jaws.

Neglect —- The words neglect, negligence, and negligently import a want of such attention to the nature or probable consequences of the act or omission as a prudent man ordinarily bestows in acting in his own concerns.

Neuter —- an animal that is neutral in the production area. Can not produce offspring.

Occiput —- The bony knob at the top of the skull between the ears

Open bitch —- A bitch that can be bred.

Overshot —- Having the lower jaw so short that the upper and lower incisors fail to meet.

Pastern —- That part of the foreleg between the pastern joint and the foot.

Pedigree —- The written record of a dog's descent (lineage) of three generations or more.

Period of gestation —- The duration of pregnancy, 63 days in the dog.

Police dog —- Any dog trained and used for police work.

Puppy —- A dog under the age of one.

Sable —- A lacing of black hairs over a lighter ground color.

Scissors bite —- a bite in which the incisors of the upper jaw just overlap and play upon those of the lower jaw.

Sire—- The male parent

Sled dogs —- Any dog trained and used to pull sledges.

Spay —- To render a bitch sterile by the surgical removal of her ovaries. To make one neutral. One may also use the word neutered.

Standard —- A description of the ideal dog of a particular breed. The blueprint of that breed.

Stud dog —- A male dog used for breeding purposes.

Tail set —- How the base of the tail sets on the rump.

Testicles —- The male gonad, gland which produces spermatozoa.

Top line —- The dog's outline from just behind the withers to the tail set.

Undercoat —- A growth of short, fine hair, or pile, partly or entirely concealed by the coarser topcoat, which grows through it.

Undershot —- Having the lower incisor teeth projecting beyond the upper ones when the mouth is closed. The opposite to overshot.

Unsound —- a dog incapable of performing the functions for which it was designed.

Whelps —- Un-weaned pups.

Withers —- The part between the shoulder bones at the base of the neck, the point from which the height of a dog is usually measured.

Alsatian Shepalutes Pedigree

The following list of pure-bred dogs have continually bred into the Alsatian Shepalute line great personality, devotion and intelligence as well as the physical beauty necessary that filtered out into this one great companion dog breed the Alsatian Shepalute! Without the absolutely wonderful characters of these purebred dogs I could not have gotten this absolutely loving and incredibly intelligent new breed of dog!

The German Shepherds

I would like to tell my readers that of all the many German Shepherds that I have bred with and recorded the offspring of, only the following German Shepherds bred great characters, soundness with the mild temperaments, consistently. These characters were so necessary in stabilizing this new breed of companion dog. I have listed them in order of the greatest concentration of genes within this new breed. The first listed having a greater impact than that of the last. As I have stated, many other shepherds were used but no other German Shepherd Dogs produced such outstanding pups, as these did!

You Can Call Me Al
Housmekon Covy Tucker Hill N.Y.
Jeff v. Flamings-Sand schH2 FH
Fanny vom perf-Gansbacheck Sch H2
King of Southaus glory
Lucca Anja Boran

The Malamutes
Sheba of Silver frost
Manfred
Cherokee chief

The English Mastiffs
Ch. Brite Star Sir Winston
Commander Cody's Sara Lee
Ch. Iron hills War Wagon

Many Thanks go to these wonderful dogs!

Bibliography

Books
The breeds and standards as recognized by the American Kennel Club

American Kennel Club. The Complete Dog Book. Howell Book House. 1985.

Balch, James F., M.D. Phyllis a Balch, C.N.C. Prescription for Nutritional Healing: a practical A to Z Reference to Drug-Free Remedies Using Vitamins, Minerals, Herbs & Food Supplements. Avery Publishing Company. 1933.

Barett, Jim. Quick Guide: Fences and Gates Creative Homeowners Press. 1998.

Barrie, Anmarie, ESQ. Dogs and the Law. t.f.h Publications. 1990.

Bennett, Jane G. The New Complete German Shepherd Dog. Howell Book House, Inc. 1988.

Covington-Thorne, Martha. Handling Your Own Dog. Doubleday and Company, Inc. 1979.

Encyclopeadia Britannica. The New Encyclopeadia Britannica. Encyclopeadia Britannica, Inc. 1986.

Glover, Harry. A Standard Guide to Pure-bred Dogs. McGraw-Hill Book Company. 1977.

Kern, Francis G. German Shepherd Dogs. T.F.H. Publications, Inc. Ltd. 1979

Landau, Sidney I. Standard Desk Dictionary. Funk and Wagnall's Publishing Company. 1974.

"Laws Pertaining to Animals" Lesson 30.

North American School of Animal Sciences. 1989.

Lopez, Barry Holstun. Of Wolves and Men. Charles Scribner's Sons. 1978.

Ricker, Elizabeth M. Seppala: Alaska Dog Driver. Little, Brown, and Company. 1930.

Riddle, Maxwell; Eva B. Seeley. The Complete Alaskan Malamute. Howell Book House, Inc. 1988.

Whitney, Leon F., D.V.M. How to Breed Dogs. Howell Book House, Inc. 1978.

Whitney, Leon F., D.V.M. Dollars in Dogs. t.f.h Publications. 1971.

Wood, Carl P. The Gun Digest Book Of Sporting Dogs

Web sites
A.O.L. search for dog breed registry clubs
widdershinestates.jwdhs-nt.com/WE/alaskan-klee-kai/origin.htm
httn://www.hund.ch/rasse/ceskvterrieruk.htm
www.fci.he/aboutus.asn?lang=eng
www.rarebreed.com/breedlist.html
www.canadianeskimodog.com/history.htm
www.cabelasiditarod.com/runyan-husky.html
www.sitstay.com/links/national-clubs/
realsgs.infor/GSDinfo/papers/intdogorgs.htm
www.woderpuppy.net/kc.htm
www.npwrc.usgs.gov/resource/distr/others.endanger/canilupu.htm
www.kats-korner.com/graywolf.html
www.boomerwolf.com/graycors.htm
www.fws.gov/r3pao/wolf/wolfindx.html

Lois Denny

www.hillsbourough.k12.nj.us/hhs/endspeci/canislupus.html
www.netpets.com/dogs/newsroom/akcfss1.html
http//outpostnet.com/abo/
americanbulldog.org/modernab.htm

About the Author

I do not write books for a living. I have never professed to a knowledge of the English language or to the exact formula on how to write a book. I am sure there must be a formula! My work and my love is with these dogs. That is what I know and that is what I do, so I ask you to go easy on me as I try to put all this information together and create this, my first book.

I sold my first trained dog at the age of 13. A mother of a 12-year-old boy gave me $50.00 and promised to pay the rest later. (I guess I don't need to tell you how that went.)

I started breeding animals at the age of 7 or 8 since then I have had the following animals share my life with me:

20 guinea pigs, a few rabbits, lots of chickens, and a bunch of ducks and geese. Horses that we got from the rescue in Nevada. Calf's, pigs, turkeys, pheasants and quail. A flock of sheep once. Love the goats and still have some. One time a couple of gerbils. Turtles whenever I could get them.

I built a couple of aviaries in my life so I had many different kinds of birds that I bred for colors such as: 50 cockatiels, maybe 100 budgies, 1 Amazon parrot, 1 African gray, a few love birds, loads of different kinds of finches and some canaries.

The different dog breeds that I have had included:

1 pointer for about 4 months, 1 German pointer until I found that one a home, a couple of labs in my life, a golden retriever for the blind in 4-H, an Irish setter until I found it a home, Cocker Spaniels (at least 100), 1 English cocker, 1 vizsla, a weimaraner (until I found it a home), basset hounds, 1 beagle that I gave to my son, dachshunds, 1 Norwegian elkhound, 4 akita's that I bred with shepherds and sold the pups, a few Alaskan Malamutes, 1 boxer, a collic, a few Dobermans, a bunch of German Shepherd Dogs that I co-owned or found homes for, Mastiffs that I bred and showed, Rottwieler's that I raised and trained for sale, 1 Samoyed that got lost, Shetlands, Siberian's that I sold, Miniature schnauzers that I sold, Scottish terriers that I sold, 1 westie that I showed and then traded for a Breeding Rottwieler, Chihuahuas that I bred and

sold, a Maltese, a couple of miniature pinschers, Pekingese, poms, poodles, pugs, shihtzus, silky t's, yorkies, a Bichon Frise, Boston terriers, bulldogs, chows, a Dalmatian, keeshond, lhasa's, aussies, a cattledog, and several mutts.

Of all the different breeds I have had I have either sold or found homes for. Some of my dogs did end up in the pound, but very few.

I taught dog obedience classes for over 20 years at naval bases, recreation centers, and colleges.

I taught grooming classes, assistant veterinarian courses along with a genetics night class on how to breed dogs.

I showed and participated in national dog clubs throughout the state of California and Oregon.

I have lived in the mountains without electricity for over 12 years of my life and I must say they were the best years of my life. There one can live at piece with nature. Anyone, who gets the opportunity to do that, I strongly suggest go!

I did most of my early breeding and trainings on this new breed in the Los Padres' National Forest.

My work is not over by a long shot! This breed has just begun and I wish to share them with the world. The breed begets itself, or reproduces itself consistently, and the pups fit the breed standards. So, the breed's character is set. Of course no pup is perfect when it comes to a breeders point of view. That is why my work is not done. Every good breeder will consistently strive to breed better than the last, towards the standards of the breed.

I, of course, believe in the character and the gentleness of these dogs as well as in their intelligence. Many new Alsatian Shepalute owners have told me that they were so happy to have found such a dog! That to me is worth its weight in gold. To be able to help or please another human being and maybe make a difference in their lives makes me very happy. To those owners I say "No, thank you!"

I wish to tell you, the reader, that I tried my best to write this book and to do the breed justice. I also hope that you the reader will enjoy my writing and will gain in life a little bit more knowledge than perhaps you may not have known. God Bless.

L. E. Schwarz